SpringerBriefs in Computer Science

More information about this series at http://www.springer.com/series/10028

Leonardo Jiménez Rodríguez • Nghi Tran
Tho Le-Ngoc

Amplify-and-Forward Relaying in Wireless Communications

Springer

Leonardo Jiménez Rodríguez
Department of Electrical and
Computer Engineering
McGill University
Montreal, QC, Canada

Nghi Tran
Department of Electrical and
Computer Engineering
The University of Akron
Akron, OH, USA

Tho Le-Ngoc
Department of Electrical and
Computer Engineering
McGill University
Montreal, QC, Canada

ISSN 2191-5768 ISSN 2191-5776 (electronic)
SpringerBriefs in Computer Science
ISBN 978-3-319-17980-3 ISBN 978-3-319-17981-0 (eBook)
DOI 10.1007/978-3-319-17981-0

Library of Congress Control Number: 2015936701

Springer Cham Heidelberg New York Dordrecht London

Printed on acid-free paper

Springer International Publishing AG Switzerland is part of Springer Science+Business Media (www.springer.com)

To our families

Preface

Relaying techniques, in which a source node communicates to a destination node with the help of a relay, have been proposed as a cost-effective solution to address the increasing demand for high data rates and reliable services over the air. As such, it is crucial to design relay systems that are able to not only provide high spectral efficiency, but also fully exploit the diversity of the relay channel. With this objective in mind, this brief aims to report on recent advances on achievable rates, power allocation schemes, and error performance for half-duplex (HD) and full-duplex (FD) amplify-and-forward (AF) single-relay systems. First, assuming the relay operates in HD mode, we discuss the capacity and respective optimal power allocation for a wide range of AF protocols over static and fading channels. Then, optimal amplification coefficients in terms of achievable rate are presented. Turning our attention to the performance with finite constellations, the error and diversity performance of AF systems are also discussed. Finally, the capacity and error performance analysis is extended to the FD relay mode of operation, where the residual self-interference due to FD transmission is explicitly taken into account.

The target audience of this Springer Brief is researchers and professionals working on current and next-generation wireless systems. The content is also valuable for advanced students interested in wireless communications and signal processing for communications.

We would like to acknowledge the financial support from the Natural Sciences and Engineering Research Council of Canada (NSERC) through various research grants.

Montréal, QC, Canada — Leonardo Jiménez Rodríguez
Akron, OH, USA — Nghi Tran
Montréal, QC, Canada — Tho Le-Ngoc
March 2015

Contents

1 Relaying: An Overview 1
1.1 Half-Duplex Relaying 1
1.2 Full-Duplex Relaying 3
1.3 Relay Functions 4
1.4 Organization of this Brief 5
1.5 Notation 6
References 7

2 Relay Protocols 11
2.1 General System Model 11
2.2 HD Protocols 12
2.2.1 NAF Protocol 12
2.2.2 DT Protocol 14
2.2.3 OAF Protocol 14
2.2.4 DHAF Protocol 15
2.2.5 TWAF Protocol 15
2.3 FD Protocols 16
2.3.1 LR Protocol 16
2.3.2 DHAF Protocol 18
2.4 Concluding Remarks 19
References 19

3 Half-Duplex AF Relaying: Capacity and Power Allocation Over Static Channels 21
3.1 Problem Formulation 22
3.2 Capacity Under Individual Power Constraints 23
3.3 Capacity Under Joint Power Constraints 28
3.4 High and Low Power Regions 31
3.4.1 Per-Node Power Constraints 31
3.4.2 Global Power Constraint 32

3.5 Illustrative Examples ... 33
3.5.1 Symmetric Network Model ... 34
3.5.2 Linear Network Model ... 35
3.6 Concluding Remarks ... 38
References ... 38

4 Half-Duplex AF Relaying: Achievable Rate and Power Allocation Over Fading Channels ... 41
4.1 Problem Formulation ... 42
4.2 Achievable Rates and Closed-Form Approximations ... 44
4.2.1 OW DHAF Systems ... 45
4.2.2 OW Cooperative Systems ... 47
4.2.3 TW Systems ... 49
4.2.4 Remarks ... 50
4.3 Optimal Power Allocation ... 51
4.3.1 DHAF Systems ... 52
4.3.2 Cooperative Systems ... 52
4.3.3 TW Systems ... 53
4.4 Closed-Form Solution for DHAF System with CI Coefficient ... 54
4.5 Illustrative Examples ... 55
4.5.1 Tightness of the Proposed Approximations ... 56
4.5.2 Optimal Power Allocation ... 57
4.6 Concluding Remarks ... 61
References ... 61

5 Half-Duplex AF Relaying: Adaptation Policies ... 63
5.1 Problem Formulation ... 64
5.2 Optimal Power Adaptation Schemes ... 65
5.2.1 OAF System ... 66
5.2.2 NAF System ... 67
5.2.3 TW System ... 70
5.3 Illustrative Examples ... 71
5.4 Concluding Remarks ... 73
References ... 73

6 Half-Duplex AF Relaying: Error Performance and Precoding ... 75
6.1 System Model ... 76
6.2 Performance Analysis ... 78
6.3 Diversity Benefits in Error-Floor Regions ... 81
6.3.1 Diversity and Coding Gain Functions ... 81
6.3.2 Optimal Class of Precoders ... 83
6.3.3 Design of Superposition Angles for Independent Mappings ... 85
6.4 Near-Capacity Design in Turbo Pinch-Off Areas ... 86
6.4.1 Multi-D Mapping in Precoded Multiple Cooperative Frames ... 87

6.4.2 EXIT Chart Analysis 88
6.5 Illustrative Examples 90
6.6 Concluding Remarks 92
References 93

7 Full-Duplex AF Relaying: Capacity Under Residual Self-Interference 95
7.1 Problem Formulation 96
7.2 Capacity and Optimal Power Allocation 97
7.3 Asymptotic Analysis 98
7.3.1 Large Source Power, Fixed Relay Power 98
7.3.2 Large Relay Power, Fixed Source Power 99
7.3.3 Large Source and Relay Power 99
7.3.4 Large Global Power 99
7.4 Illustrative Examples and Comparisons to HD Schemes 100
7.5 Concluding Remarks 102
References 102

8 Full-Duplex AF Relaying: Error Performance Under Residual Self-Interference 105
8.1 System Model 106
8.2 PEP Analysis and Tight BER Bounds 107
8.2.1 PEP for the LR System 108
8.2.2 PEP for the DH System 110
8.2.3 Tight BER Bounds of BICM-ID Systems 111
8.3 Diversity and Coding Gain Analysis 112
8.3.1 LR Analysis 112
8.3.2 DH Analysis 115
8.4 Illustrative Examples 117
8.4.1 LR Examples 117
8.4.2 DH Examples 119
8.5 Concluding Remarks 120
References 121

Acronyms

3GPP	Third Generation Partnership Project
AF	Amplify-and-forward
AWGN	Additive white Gaussian noise
BER	Bit error rate
BF	Beamforming
BICM	Bit-interleaved coded modulation
BICM-ID	Bit-interleaved coded modulation with iterative decoding
CDI	Channel distribution information
CF	Compress-and-forward
CI	Channel inversion
CSI	Channel state information
dB	Decibels
DF	Decode-and-forward
DH	Dual-hop
DHAF	Dual-hop amplify-and-forward
DT	Direct transmission
EXIT	Extrinsic information transfer
FD	Full-duplex
FG	Fixed-gain
FR	Full-rank
HD	Half-duplex
KKT	Karush–Kuhn–Tucker
LLR	Log-likelihood ratio
LR	Linear relaying
LTE	Long Term Evolution
MAP	Maximum *a posteriori* probability
MIMO	Multiple-input multiple-output
ML	Maximum likelihood
MRC	Maximal-ratio combining
multi-D	Multidimensional
NAF	Non-orthogonal amplify-and-forward

OAF	Orthogonal amplify-and-forward
OW	One-way
PEP	Pairwise error probability
QAM	Quadrature amplitude modulation
QPSK	Quadrature phase-shift keying
RA	Relay adaptation
SISO	Soft-input soft-output
SNR	Signal-to-noise ratio
TW	Two-way
TWAF	Two-way amplify-and-forward
VG	Variable-gain

Chapter 1
Relaying: An Overview

Current wireless communications systems face increasing challenges due to the ever growing demand for high data rates and reliable services over the air. As new applications for small wireless devices arise, such rates are expected to be attained using as low power consumption as possible. In addition, bandwidth is a scarce resource and thus wireless systems are also expected to transmit high data rates in a spectrally-efficient manner. All of these requirements need to be considered in the design of emerging and future generations of wireless networks.

Relaying techniques, in which helper nodes aid in transmission from one node to another, have been recently proposed as a cost-effective solution to meet some of the demands in next generations of wireless systems [20, 27, 30, 38]. In particular, in the context of cellular networks, the deployment of relay nodes has been shown to extend and/or improve the coverage, enhance the reliability, and improve the spectral efficiency per unit area. This can be achieved without incurring the associated high costs of adding extra base stations, e.g., site acquisition and backhaul costs. As such, relaying is one of the key features currently being considered in several wireless standards such as the Third Generation Partnership Project (3GPP) Long Term Evolution (LTE), among others [27, 38]. Consequently, it is important for future wireless standards to have relay schemes that not only increase the reliability of the wireless network, but also present a high spectral efficiency [10, 29].

1.1 Half-Duplex Relaying

The relay channel, in which a source node communicates to a destination node with the help of a relay, was first introduced by van der Meulen in [36]. Earlier information theoretical works assumed that the relay is capable of operating in

L. Jiménez Rodríguez et al., *Amplify-and-Forward Relaying in Wireless Communications*, SpringerBriefs in Computer Science,
DOI 10.1007/978-3-319-17981-0_1

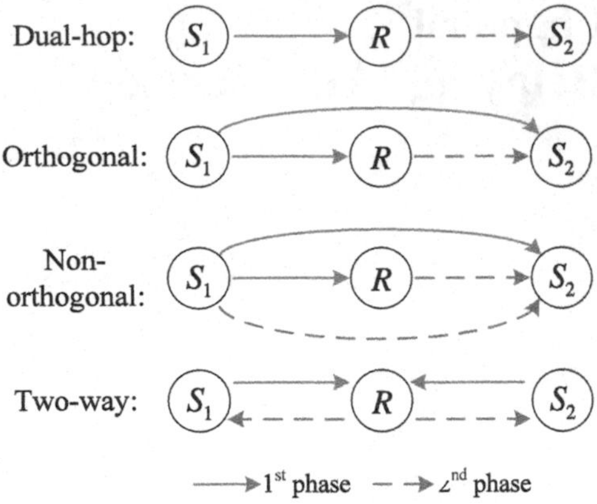

Fig. 1.1 HD relay protocols

full-duplex (FD) mode, i.e., the relay is able to transmit and receive at the same time and over the same frequency band [8, 19, 24, 36, 37]. This assumption was generally believed to be impractical due to the great difference in transmit and receive signal powers levels, which results in self-interference. Thus, motivated by wireless scenarios, the focus on the relay channel was shifted to *half-duplex* (HD) operation.

Seminal works on HD relaying concentrated on the *dual-hop* (DH) strategy, e.g., [6, 15–18]. In DH protocols, information is transmitted from source to destination via the relay in two phases as shown in Fig. 1.1. In the first phase, the source node S_1 transmits to the relay node R, whereas in the second phase, the relay communicates to the destination node S_2 and the source remains silent. Such a DH strategy can be easily implemented in practice to improve the coverage of the network and has already been considered for next-generation mobile communication standards [27, 38]. Furthermore, DH schemes are the only alternative when the *direct* link from source to destination is under severe shadowing. However, when the direct link is available for transmission, DH techniques present two main limitations. First, due to the HD constraint, DH systems are not spectrally-efficient as the source is only allowed to transmit every other phase. Second, although path-loss savings can be obtained, DH protocols provide no diversity gain as there exists only one path from source to destination.

By considering the direct source-destination link, *cooperative* relaying protocols have been introduced to counteract the drawbacks of DH relaying when the direct link is available [2, 25, 26, 28, 29, 35]. Pioneering works on cooperative relaying focused on *orthogonal* schemes in which the source and relay alternate for transmission as in DH systems. However, contrary to DH schemes, the destination also listens to the source node in the first phase as shown in Fig. 1.1. Two independent paths between the source and the destination (the direct and the relay link) are thus created. With proper combining at the destination, rate and diversity advantages over DH protocols can be obtained. This diversity gain, commonly referred to as *cooperative diversity*, is particularly important for small devices which cannot accommodate multiple antennas due to practical constraints. In this case, spatial diversity can still be realized through cooperative relaying in a distributed fashion.

Although orthogonal schemes provide rate and diversity advantages compared to DH transmission, the spectral efficiency of these cooperative protocols still suffers from the HD constraint of the relay. This is because similar to DH relaying, the source must remain silent in the second transmission phase. To mitigate the impact of HD relaying, *non-orthogonal* schemes in which the source is allowed to transmit continuously have been proposed in the literature [2, 28]. In non-orthogonal protocols, as illustrated in Fig. 1.1, the source and relay transmit concurrently in the second phase, maximizing the degrees of broadcasting and receive collision [28]. Moreover, non-orthogonal protocols are general in that they include orthogonal schemes and even direct transmission (i.e., when the relay is not used) as special cases. Unfortunately, the analysis and optimization of non-orthogonal networks are more challenging than their orthogonal counterparts and therefore have received less attention in the literature [2, 9, 28].

In the protocols discussed above, the node S_1 wants to communicate to S_2 via the relay, i.e., information flows from S_1 to S_2 ($S_1 \rightarrow S_2$). These protocols can thus be broadly classified as *one-way* (OW) relaying schemes. On the other hand, several applications require the nodes S_1 and S_2 to exchange information in a bidirectional or *two-way* (TW) fashion. A simple method to achieve this is by applying any of the above OW protocols, i.e., $S_1 \rightarrow S_2$ over the first two phases and $S_2 \rightarrow S_1$ over the remaining two. This exchange would require four transmission phases and only two symbols would be exchanged if we were to use orthogonal or DH schemes. To mitigate the impact of HD relaying for bidirectional communication, two-phase and the three-phase TW protocols have recently been proposed in the literature [23, 32]. Specifically, in the two-phase scheme, both source nodes simultaneously communicate to the relay in the first phase, whereas the relay broadcasts in the second phase (see Fig. 1.1). In the three-phase scheme, the source nodes take turns to communicate to the relay in the first two phases and the relay broadcasts in the third. Similar to OW non-orthogonal protocols, the potential benefits of TW protocols have just started to be investigated.

1.2 Full-Duplex Relaying

It can be seen from the above discussion that the HD constraint of the relay has a great impact on the spectral efficiency of relaying protocols. As explained before, this HD constraint is motivated by the fact that the transmit signal power is usually orders of magnitude larger than the received signal power, resulting in heavy self-interference. Although originally believed to be impractical, FD wireless operation has been recently shown to be feasible through novel combinations of self-interference mitigation schemes (see for example [1, 4, 5, 7, 11–14, 21, 22, 31, 33, 34] and references within). In particular, to avoid saturating the receiver

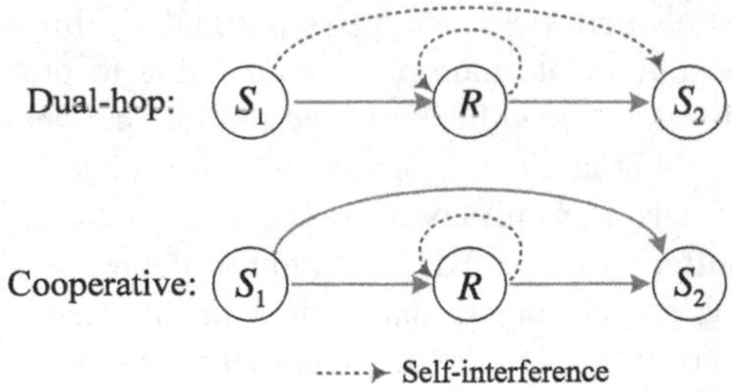

Fig. 1.2 FD relay protocols

front end, several techniques prior to analog-to-digital conversion have been proposed. For instance, basic analog cancellation methods include antenna separation [11, 12, 14, 33], orientation [33, 34] and directionality [13, 31]. More involved methods include asymmetric placement of transmit antennas [7], symmetric placement of antennas with phase shifters [1, 22], the use of a balanced/unbalanced transformer [21], the use of a circulator [4], and analog time domain subtraction [4, 11–13, 31, 34], among others. These analog techniques can be combined with digital methods after quantization such as time domain subtraction for further mitigation [4, 11–13, 21]. Despite these advances in cancellation techniques, the self-interference remains a challenge as it cannot be completely mitigated in practice. As such, different from earlier information theoretical works, the *residual self-interference* must be explicitly taken into account when assessing, designing and analyzing FD protocols.

Similar to HD schemes, FD protocols can be classified depending on whether the direct source-destination link is used for transmission. In fact, the idea of cooperative relaying can be traced back to the works of van der Meulen and Cover in [8, 36] which make use of the direct link for FD communication. Specifically, in cooperative FD schemes, the source transmits to the relay and the destination, while the relay simultaneously receives the signal from the source and transmits to the destination, as shown in Fig. 1.2. Analogous to the HD DH scheme, the direct link is not used in FD DH protocols. Similar to their HD counterparts, FD DH relaying has two main limitations when the direct link is not under heavy shadowing. First, although the source is allowed to transmit continuously, the rate of the FD DH scheme might be degraded due to the self-interference created at the destination node from the direct link. Furthermore, this protocol does not provide any diversity benefits. Thus, as in the HD scenario, cooperative FD techniques that make use of the direct link for transmission might be able to provide rate and diversity advantages.

1.3 Relay Functions

To this day, the capacity of the general relay channel along with its respective optimal relay function are unknown. Thus, several relay functions have been proposed in the literature. Common relay functions include *decode-and-forward*

(DF), *compress-and-forward* (CF), and *amplify-and-forward* (AF). In DF, the relay first decodes the information received from the source node, encodes it and then forwards it to the destination node [8]. On the other hand, in CF, the relay forwards a quantized version of the received signal to the destination [8]. Finally, the relay simply amplifies the signal received in the previous phase and forwards it to the destination in AF [26]. Among these three strategies, the AF technique is of practical interest as it requires lower implementation and computational complexity, it carries less delay at the relay terminal, and it is transparent to the modulation/coding used by the source nodes [3]. As such, the focus of this brief is on AF relay protocols.

AF relaying can be further classified according to the availability of channel side information (CSI) at the relay node. In particular, to maintain a long-term average power constraint at the relay, most previous studies on AF relaying considered two power amplification techniques. The first method assumes that the relay has only channel distribution information (CDI) of the source-relay link and amplifies the received signal using a *fixed-gain* (FG) coefficient [28]. The second technique assumes that the relay has an instantaneous knowledge of the source-relay channel and uses this channel gain to normalize the received signal to the desired power level [26]. The latter variable-gain (VG) method is therefore referred to as the *channel inversion* (CI) coefficient.

1.4 Organization of this Brief

As noted before, future wireless networks require relaying protocols that are able to fully exploit the diversity of the channel as well as provide high spectral efficiency. Motivated by this fact, the objective of this brief is to report on recent advances on achievable rates, power allocation schemes, and error performance of HD and FD AF relay systems. The outline of this brief is as follows.

First, Chap. 2 introduces the input-output relations of the considered relay protocols.

Then, Chap. 3 discusses the capacity of the static HD non-orthogonal AF (NAF) channel under both per-node and joint power constraints. In particular, by deriving and comparing all local solutions, the optimal input covariance matrix at the source and the optimal power allocation scheme at the relay are characterized. The capacity of the NAF system is also analyzed for some concrete examples, such as under different transmission power levels and several network models.

Considering Rayleigh fading channels, Chap. 4 presents a general method to analyze the achievable rate and to characterize the optimal power allocation scheme for a wide range of HD AF protocols. By exploiting the capacity of a two-branch maximal-ratio combining (MRC) system and a simple approximation to the logarithm, tight yet simple approximations to the achievable rates are obtained in high and low transmission power regimes. Then, using the derived approximations, the FG and CI coefficients are compared. In addition, we analytically quantify the asymptotic power allocation schemes among the nodes to achieve the maximum rate and sum rate for OW and TW schemes, respectively.

As the CI and FG coefficients are simply power normalization factors, Chap. 5 develops optimal amplification coefficients at the relay for HD cooperative and TW AF systems. The achievable rate and sum rate criteria for OW and TW relaying are considered, respectively. In particular, assuming full knowledge of all channel gains at the relay and Gaussian inputs at the source, optimal power adaptation schemes under a long-term average power constraint are established in closed form.

Turning our attention to the performance with finite constellations, in Chap. 6 we propose the idea of precoding over multiple cooperative frames to improve the diversity and approach the capacity of a HD NAF system over Rayleigh fading channels. At first, a tight union bound on the bit error rate (BER) is derived by adopting the coding framework of bit-interleaved coded modulation (BICM). Focusing on the error-floor region, we then show that a significantly higher diversity order can be achieved by precoding over multiple frames. Subsequently, concentrating on the turbo pinch-off region, we demonstrate that a concatenation of multi-dimensional (multi-D) mapping and multiple-frame precoding can be used to approach the capacity of the HD NAF channel.

Concentrating on the FD mode of operation, Chap. 7 investigates the capacity limit of a FD DH system over static channel gains. Assuming that the residual self-interference variance is proportional to the λ-th power of the transmitted power, the optimal allocation schemes are derived under both per-node and sum power constraints. The capacity and optimal schemes are then analyzed in different high power regions. Comparisons between the HD and FD systems are also carried out.

Still considering FD transmission and under the same interference model as above, Chap. 8 investigates the error and diversity performance of FD AF systems over Rayleigh fading channels. The study focuses on the cooperative linear relaying (LR) protocol with direct source-destination link and the DH scheme without direct link, both under uncoded and coded frameworks. At first, closed-form pairwise error probability (PEP) expressions are derived for the uncoded systems, which are then used to obtain tight bounds to the BER of the coded systems. To shed an insight on the diversity behavior, asymptotic expressions at high transmission powers are then presented.

1.5 Notation

In this brief, matrices and vectors are denoted with bold upper- and lower-case letters, respectively. The identity matrix of size $N \times N$ is denoted as $\boldsymbol{I}_N$ and the zero vector of appropriate size is given by $\mathbf{0}$. The transpose, Hermitian, determinant and trace operators are respectively denoted as $^\top$, †, $\det(\cdot)$ and $\mathrm{tr}(\cdot)$. The function $\mathrm{diag}(\boldsymbol{a})$ returns a diagonal matrix with the vector $\boldsymbol{a}$ in the diagonal, and $\mathrm{bdiag}(\boldsymbol{A}_1, \ldots, \boldsymbol{A}_N)$ is a block diagonal matrix with diagonal components $\boldsymbol{A}_i$. A positive semidefinite (psd) matrix $\boldsymbol{A}$ is denoted as $\boldsymbol{A} \succeq \boldsymbol{0}$. A zero-mean circularly Gaussian distributed vector with covariance matrix $\boldsymbol{A}$ is denoted as $\mathscr{CN}(\boldsymbol{0}, \boldsymbol{A})$. $\mathbb{E}[\cdot]$ is the expectation operator. Finally, $\log(\cdot)$ and $\ln(\cdot)$ are the base-two and natural logarithms.

References

1. Aryafar, E., Khojastepour, M.A., Sundaresan, K., Rangarajan, S., Chiang, M.: MIDU: enabling MIMO full-duplex. In: Proceedings of ACM annual international conference on Mobile computing networks, pp. 257–268. (2012). Doi:10.1145/2348543.2348576
2. Azarian, K., Gamal, H.E., Schniter, P.: On the achievable diversity-multiplexing tradeoff in half-duplex cooperative channels. IEEE Trans. Inf. Theory **51**(12), 4152–4172 (2005). Doi:10.1109/TIT.2005.858920
3. Berger, S., Kuhn, M., Wittneben, A., Unger, T., Klein, A.: Recent advances in amplify-and-forward two-hop relaying. IEEE Commun. Mag. **47**(7), 50–56 (2009). Doi:10.1109/MCOM.2009.5183472
4. Bharadia, D., McMilin, E., Katti, S.: Full-duplex radios. In: Proc. ACM SIGCOMM, pp. 375–386 (2013). Doi:10.1145/2486001.2486033
5. Bliss, D., Parker, P., Margetts, A.: Simultaneous transmission and reception for improved wireless network performance. In: IEEE Statistical Signal Processing Workshop, pp. 478–482 (2007). Doi:10.1109/SSP.2007.4301304
6. Bölcskei, H., Nabar, R., Oyman, O., Paulraj, A.: Capacity scaling laws in MIMO relay networks. IEEE Trans. Wireless Commun. **5**(6), 1433–1444 (2006). Doi:10.1109/TWC.2006.1638664
7. Choi, J.I., Jain, M., Srinivasan, K., Levis, P., Katti, S.: Achieving single channel, full-duplex wireless communication. In: Proceedings of ACM Annual International Conference Mobile Computing Network, pp. 1–12 (2010). Doi:10.1145/1859995.1859997
8. Cover, T., Gamal, A.: Capacity theorems for the relay channel. IEEE Trans. Inf. Theory **25**(5), 572–584 (1979). Doi:10.1109/TIT.1979.1056084
9. Ding, Y., Zhang, J.K., Wong, K.M.: Ergodic channel capacities for the amplify-and-forward half-duplex cooperative systems. IEEE Trans. Inf. Theory **55**(2), 713–730 (2009). Doi:10.1109/TIT.2008.2009822
10. Ding, Z., Krikidis, I., Rong, B., Thompson, J., Wang, C., Yang, S.: On combating the half-duplex constraint in modern cooperative networks: protocols and techniques. IEEE Wireless Commun. Mag. **19**(6), 20–27 (2012). Doi:10.1109/MWC.2012.6393514
11. Duarte, M., Sabharwal, A.: Full-duplex wireless communications using off-the-shelf radios: Feasibility and first results. In: Proceedings of Asilomar Conference on Signals, System, Computers, pp. 1558–1562 (2010). Doi:10.1109/ACSSC.2010.5757799
12. Duarte, M., Dick, C., Sabharwal, A.: Experiment-driven characterization of full-duplex wireless systems. IEEE Trans. Wireless Commun. **11**(12), 4296–4307 (2012). Doi:10.1109/TWC.2012.102612.111278
13. Everett, E., Duarte, M., Dick, C., Sabharwal, A.: Empowering full-duplex wireless communication by exploiting directional diversity. In: Proceeding of Asilomar Conference on Signals, System, Computers, pp. 2002–2006 (2011). Doi:10.1109/ACSSC.2011.6190376
14. Haneda, K., Kahra, E., Wyne, S., Icheln, C., Vainikainen, P.: Measurement of loop-back interference channels for outdoor-to-indoor full-duplex radio relays. In: Proceedings of European Conference Antenna Propagation, pp. 1–5 (2010)
15. Hasna, M., Alouini, M.S.: End-to-end performance of transmission systems with relays over Rayleigh-fading channels. IEEE Trans. Wireless Commun. **2**(6), 1126–1131 (2003). Doi:10.1109/TWC.2003.819030

16. Hasna, M., Alouini, M.S.: Outage probability of multihop transmission over Nakagami fading channels. IEEE Commun. Lett. **7**(5), 216–218 (2003). Doi:10.1109/LCOMM.2003.812178
17. Hasna, M., Alouini, M.S.: A performance study of dual-hop transmissions with fixed gain relays. IEEE Trans. Wireless Commun. **3**(6), 1963–1968 (2004). Doi:10.1109/TWC.2004.837470
18. Hasna, M., Alouini, M.S.: Optimal power allocation for relayed transmissions over Rayleigh-fading channels. IEEE Trans. Wireless Commun. **3**(6), 1999–2004 (2004). Doi:10.1109/TWC.2004.833447
19. Høst-Madsen, A., Zhang, J.: Capacity bounds and power allocation for wireless relay channels. IEEE Trans. Inf. Theory **51**(6), 2020–2040 (2005). Doi:10.1109/TIT.2005.847703
20. Hoymann, C., Chen, W., Montojo, J., Golitschek, A., Koutsimanis, C., Shen, X.: Relaying operation in 3GPP LTE: challenges and solutions. IEEE Commun. Mag. **50**(2), 156–162 (2012). Doi:10.1109/MCOM.2012.6146495
21. Jain, M., Choi, J.I., Kim, T., Bharadia, D., Seth, S., Srinivasan, K., Levis, P., Katti, S., Sinha, P.: Practical, real-time, full-duplex wireless. In: Proceedings of ACM Annual International Conference on Mobile Computing, pp. 301–312 (2011). Doi:10.1145/2030613.2030647
22. Khojastepour, M.A., Sundaresan, K., Rangarajan, S., Zhang, X., Barghi, S.: The case for antenna cancellation for scalable full-duplex wireless communications. In: Proceedings of ACM Workshop Hot Topics Network, pp. 17:1–17:6 (2011). Doi:10.1145/2070562.2070579
23. Kim, S.J., Devroye, N., Mitran, P., Tarokh, V.: Achievable rate regions and performance comparison of half-duplex bi-directional relaying protocols. IEEE Trans. Inf. Theory **57**(10), 6405–6418 (2011). Doi:10.1109/TIT.2011.2165132
24. Kramer, G., Gastpar, M., Gupta, P.: Cooperative strategies and capacity theorems for relay networks. IEEE Trans. Inf. Theory **51**(9), 3037–3063 (2005). Doi:10.1109/TIT.2005.853304
25. Laneman, J.N., Wornell, G.W.: Distributed space-time-coded protocols for exploiting cooperative diversity. IEEE Trans. Inf. Theory **49**(10), 2415–2425 (2003). Doi:10.1109/TIT.2003.817829
26. Laneman, J., Tse, D., Wornell, G.: Cooperative diversity in wireless networks: Efficient protocols and outage behavior. IEEE Trans. Inf. Theory **50**(12), 3062–3080 (2004). Doi:10.1109/TIT.2004.838089
27. Loa, K., Wu, C.C., Sheu, S.T., Yuan, Y., Chion, M., Huo, D., Xu, L.: IMT-advanced relay standards [WiMAX/LTE update]. IEEE Commun. Mag. **48**(8), 40–48 (2010). Doi:10.1109/MCOM.2010.5534586
28. Nabar, R.U., Bölcskei, H., Kneubühler, F.W.: Fading relay channel: Performance limits and space-time signal design. IEEE J. Sel. Areas Commun. **22**(6), 1099–1109 (2004). Doi:10.1109/JSAC.2004.830922
29. Nosratinia, A., Hunter, T., Hedayat, A.: Cooperative communication in wireless networks. IEEE Commun. Mag. **42**(10), 74–80 (2004). Doi:10.1109/MCOM.2004.1341264
30. Pabst, R., Walke, B.H., Schultz, D., Herhold, P., Yanikomeroglu, H., Mukherjee, S., Viswanathan, H., Lott, M., Zirwas, W., Dohler, M., Aghvami, H., Falconer, D., Fettweis, G.: Relay-based deployment concepts for wireless and mobile broadband radio. IEEE Commun. Mag. **42**(9), 80–89 (2004). Doi:10.1109/MCOM.2004.1336724
31. Radunovic, B., Gunawardena, D., Key, P., Proutiere, A., Singhy, N., Balan, V., Dejean, G.: Rethinking indoor wireless: Low power, low frequency, full-duplex. Technical Report MSR-TR-2009-148, Microsoft Research (2009)
32. Rankov, B., Wittneben, A.: Spectral efficient protocols for half-duplex fading relay channels. IEEE J. Sel. Areas Commun. **25**(2), 379–389 (2007). Doi:10.1109/JSAC.2007.070213
33. Riihonen, T., Balakrishnan, A., Haneda, K., Wyne, S., Werner, S., Wichman, R.: Optimal eigenbeamforming for suppressing self-interference in full-duplex MIMO relays. In: Proceedings of Annual Conference on Information Science and Systems, pp. 1–6 (2011). Doi:10.1109/CISS.2011.5766241
34. Sahai, A., Patel, G., Sabharwal, A.: Pushing the limits of full-duplex: Design and real-time implementation. arXiv (2011). URL http://arxiv.org/abs/1107.0607

35. Sendonaris, A., Erkip, E., Aazhang, B.: User cooperation diversity. Part I. System description. IEEE Trans. Commun. **51**(11), 1927–1938 (2003). Doi:10.1109/TCOMM.2003.818096
36. van der Meulen, E.: Three-terminal communication channels. Adv. Appl. Probab. **3**(1), 120–154 (1971)
37. Wang, B., Zhang, J., Høst-Madsen, A.: On the capacity of MIMO relay channels. IEEE Trans. Inf. Theory **51**(1), 29–43 (2005). Doi:10.1109/TIT.2004.839487
38. Yang, Y., Hu, H., Xu, J., Mao, G.: Relay technologies for WiMax and LTE-advanced mobile systems. IEEE Commun. Mag. **47**(10), 100–105 (2009). Doi:10.1109/MCOM.2009.5273815

Chapter 2
Relay Protocols

This chapter introduces the input-output relations for the AF relay schemes considered in this brief.

2.1 General System Model

The system considered in this brief, shown in Fig. 2.1, consists of three single-antenna nodes: a relay node R, and two HD source nodes S_1 and S_2. The relay node R assists in the transmission between S_1 and S_2, and might operate in either HD or FD mode. In the FD mode, the residual self-interference is explicitly taken into account. The channel gain from S_1 to S_2 is denoted by h_0, whereas the gains from S_1 to R and S_2 to R are given respectively by h_1 and h_2. These channel gains are assumed to be reciprocal so that the gains from node A to node B and from B to A are the same, $A, B \in \{S_1, S_2, R\}$ $(A \neq B)$. For simplicity, we denote $h_l = \sqrt{\alpha_l} e^{j\theta_l}$, where α_l and θ_l are the magnitude squared and phase of h_l, $l \in \{0, 1, 2\}$. In this brief, we consider *frequency-flat Rayleigh fading* so that the channel gains are independent zero-mean circularly Gaussian distributed which is denoted as $h_l \sim \mathscr{CN}(0, \phi_l)$, i.e., with Rayleigh distributed magnitude. Furthermore, we adopt the *block fading* model such that the channel gains $\boldsymbol{h} = [h_0, h_1, h_2]$ remain constant for a coherence interval T_c (given in number of symbol periods) and change independently after. We consider both the static scenario (where transmitted codeword spans over a time-invariant $\boldsymbol{h}$) and the fast fading one (where it spans over multiple realizations of $\boldsymbol{h}$). We first introduce the HD protocols of interest before describing the FD ones.

L. Jiménez Rodríguez et al., *Amplify-and-Forward Relaying in Wireless Communications*, SpringerBriefs in Computer Science,
DOI 10.1007/978-3-319-17981-0_2

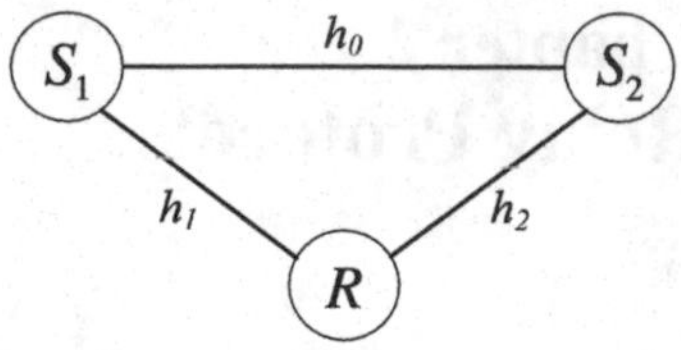

Fig. 2.1 Relay system

2.2 HD Protocols

The considered HD protocols are described in the following.

2.2.1 NAF Protocol

We first introduce the OW *non-orthogonal AF* (NAF) protocol, which is obtained by combining non-orthogonal transmission in Fig. 1.1 with the AF relaying scheme. Specifically, we consider the NAF scheme studied in [1, 3, 7], which has been shown to be optimal in terms of diversity-multiplexing tradeoff over the single-relay AF channel [1]. Without loss of generality, we assume S_1 is the source node S and S_2 the destination node D. In this protocol, transmission is carried in cooperative frames composed of two unit-time phases (where $T_c \geq 2$ is a multiple of two). In the first phase, denoted as the broadcasting phase, S sends the first signal x_1 to both R and D. The received signals at R and D can be written respectively as

$$y_{r,1} = \sqrt{P_s} h_1 x_1 + n_{r,1}, \quad \text{and} \quad y_1 = \sqrt{P_s} h_0 x_1 + n_{d,1},$$

where P_s is a constant related to the transmission power at S; and $n_{r,1}$ and $n_{d,1}$ are the independent noise samples with respective variance N_r and N_d, denoted as $n_{r,1} \sim \mathcal{CN}(0, N_r)$ and $n_{d,1} \sim \mathcal{CN}(0, N_d)$. In the second or cooperative phase, S sends the signal x_2 to D while R amplifies and forwards the symbol received during the first phase to D using an amplification coefficient b. The received signal at D in the second phase is expressed as

$$y_2 = \sqrt{P_s} h_0 x_2 + \sqrt{P_r} b h_2 y_{r,1} + n_{d,2} = \sqrt{P_s} h_0 x_2 + \sqrt{P_r} b h_2 (\sqrt{P_s} h_1 x_1 + n_{r,1}) + n_{d,2},$$

where P_r is a constant related to the transmission power at R and $n_{d,2} \sim \mathcal{CN}(0, N_d)$.

The signal received at D over these two phases, $\boldsymbol{y} = [y_1, y_2]^\top$, can be written in matrix form as in [3]:

$$y = \sqrt{P_s}\underbrace{(h_0 I_2 + \sqrt{P_r} h_1 h_2 B)}_{H_{\text{NAF}}} x + \underbrace{\sqrt{P_r} h_2 B n_r + n_d}_{n} = \sqrt{P_s} H_{\text{NAF}} x + n, \tag{2.1}$$

where $x = [x_1, x_2]^\top$ is the input vector; $B = \begin{pmatrix} 0 & 0 \\ b & 0 \end{pmatrix}$; $n_r = [n_{r,1}, n_{r,2}]^\top \sim \mathcal{CN}(\mathbf{0}, N_r I_2)$ is the noise vector at R; and $n_d = [n_{d,1}, n_{d,2}]^\top \sim \mathcal{CN}(\mathbf{0}, N_d I_2)$ is the noise vector at D. The equivalent 2×2 channel matrix in (2.1) is thus

$$H_{\text{NAF}} = \begin{pmatrix} h_0 & 0 \\ \sqrt{P_r} b h_1 h_2 & h_0 \end{pmatrix}, \tag{2.2}$$

and the equivalent noise vector $n = [n_1, n_2]^\top = [n_{d,1}, \sqrt{P_r} b h_2 n_{r,1} + n_{d,2}]^\top \sim \mathcal{CN}(\mathbf{0}, K)$ with

$$K = \begin{pmatrix} N_d & 0 \\ 0 & N_d + P_r b^2 \alpha_2 N_r \end{pmatrix}. \tag{2.3}$$

Let the temporal covariance matrix of the input vector x be denoted as $Q = \mathbb{E}[x x^\dagger] = \begin{pmatrix} q_1 & q_{12} \\ q_{12}^\dagger & q_2 \end{pmatrix} \succeq \mathbf{0}$. From the covariance matrix and (2.1), it is straightforward to see that S allocates a power of $q_1 P_s$ to x_1 in the first phase and $q_2 P_s$ to x_2 in the second phase. The average transmitted power at S is then $P_s \cdot \text{tr}(Q) = P_s(q_1 + q_2)$ per transmission frame or $(P_s/2) \cdot \text{tr}(Q) = P_s(q_1 + q_2)/2$ per symbol period. Now, let $z_2 P_r$ be the average power constraint at R in the second phase, i.e., $\mathbb{E}[b^2 |\sqrt{P_s} h_1 x_1 + n_{r,1}|^2] = z_2$. The amplification coefficient can then be expressed as

$$b = \sqrt{\frac{z_2}{q_1 P_s \Xi_1 + N_r}}, \tag{2.4}$$

where the parameter Ξ_1 depends on the CSI available at the relay. For instance, when the relay only knows the second order statistics of the S-R channel, the FG CDI coefficient is obtained by setting $\Xi_1 = \phi_1$. On the other hand, when the relay has instantaneous knowledge of the $S - R$ link, the VG CI coefficient is given with $\Xi_1 = \alpha_1$.

As a shorthand notation, let the mutual information between the input x and output y vectors of (2.1) conditioned on a realization of h be given by $I|h$. Assuming Gaussian codebooks at the source, the conditional achievable rate (in b/s/Hz) can be written from (2.1) as in [3]:

$$I|h = \frac{1}{2} \log \det(I_2 + P_s H_{\text{NAF}}^\dagger K^{-1} H_{\text{NAF}} Q), \tag{2.5}$$

with $\boldsymbol{K}$ as in (2.3) and $\boldsymbol{H}_{\text{NAF}}$ as in (2.2). In (2.5), the 1/2 pre-log factor is due to the fact that the transmission protocol is carried over two phases. The unconditional mutual information or ergodic achievable rate can then be obtained by averaging (2.5) over the channel gains, i.e., $I = \mathbb{E}[I\,|\boldsymbol{h}]$. Note that the NAF protocol introduced above is general, in the sense that it includes the direct transmission scheme and other OW relay protocols as explained below.

2.2.2 DT Protocol

In the *direct transmission* (DT) protocol, the relay is not used for transmission $b = 0$ and the source simply communicates its information through the direct S-D link. Although the DT scheme is not a relay protocol, it is an important alternative in cooperative systems where using the relay might not always be beneficial. By substituting $z_2 = 0$ in (2.1), the input-output relation for the DT scheme is given by

$$\boldsymbol{y} = \sqrt{P_s}\begin{pmatrix} h_0 & 0 \\ 0 & h_0 \end{pmatrix}\boldsymbol{x} + \boldsymbol{n}_d. \tag{2.6}$$

The conditional rate in (2.5) when x_1 and x_2 are independent ($q_{12} = 0$) and $z_2 = 0$ reduces to

$$I\,|\boldsymbol{h} = \frac{1}{2}\left[\log\left(1 + \frac{q_1\alpha_0 P_s}{N_d}\right) + \log\left(1 + \frac{q_2\alpha_0 P_s}{N_d}\right)\right]. \tag{2.7}$$

2.2.3 OAF Protocol

We now describe the *orthogonal AF* (OAF) in which the source and relay alternate for transmission. This protocol is given by combining orthogonal transmission in Fig. 1.1 with AF relaying and can be obtained by setting $q_2 = 0$ in the NAF protocol. Specifically, the input-output relation in (2.1) simplifies when $q_2 = 0$ to

$$\boldsymbol{y} = \sqrt{P_s}\boldsymbol{h}_{\text{OAF}}x_1 + \boldsymbol{n}, \tag{2.8}$$

where $\boldsymbol{h}_{\text{OAF}} = [h_0, \sqrt{P_r}bh_1h_2]^{\top}$ is the first column of (2.2) and the covariance matrix of $\boldsymbol{n}$ is still given as in (2.3). The conditional achievable rate of the OAF system is then given from (2.8) as

$$I\,|\boldsymbol{h} = \frac{1}{2}\log\left(1 + \frac{q_1\alpha_0 P_s}{N_d} + \frac{\alpha_1\alpha_2 q_1 b^2 P_s P_r}{N_d + \alpha_2 b^2 P_r N_r}\right). \tag{2.9}$$

2.2.4 DHAF Protocol

In the *dual-hop AF* (DHAF) protocol in Fig. 1.1, the direct link from source to destination is either ignored or heavily shadowed so that it cannot be used for communication. The input-output relation in this case can be obtained by setting $q_2 = 0$ and $h_0 = 0$ in (2.1) as

$$y = \sqrt{P_s P_r} b h_1 h_2 x_1 + n_2, \tag{2.10}$$

where $n_2 \sim \mathscr{CN}(0, N_d + b^2 \alpha_2 P_r N_r)$. The conditional rate of the DHAF system is then given by

$$I|\boldsymbol{h} = \frac{1}{2} \log \left(1 + \frac{\alpha_1 \alpha_2 q_1 b^2 P_s P_r}{N_d + \alpha_2 b^2 P_r N_r} \right). \tag{2.11}$$

2.2.5 TWAF Protocol

We now turn our attention to the *two-way AF* (TWAF) protocol in which S_1 and S_2 want to exchange information as in Fig. 1.1. In particular, we consider the two-phase TW system in [6, 8] as it provides a higher multiplexing gain than the three-phase scheme [5]. Transmission in this protocol is again carried in frames divided in two unit-time phases (where $T_c \geq 2$ is a multiple of two). In the first phase, also referred to as the multiple-access phase, S_1 and S_2 transmit simultaneously their respective symbols x_1 and x_2 to R. The received signal at R is given by

$$y_{r,1} = \sqrt{P_{s1}} h_1 x_1 + \sqrt{P_{s2}} h_2 x_2 + n_{r,1},$$

where $n_{r,1} \sim \mathscr{CN}(0, N_r)$, and P_{si} is constant related to the power transmitted at node S_i ($i \in \{1, 2\}$). In the second phase, denoted as the broadcasting phase, R simply amplifies the received signal and broadcasts it to S_1 and S_2. The signal received at S_i can be written as

$$y_i' = \sqrt{P_r} b h_i y_{r,1} + n_{d,i} = \sqrt{P_r} b h_i (\sqrt{P_{s1}} h_1 x_1 + \sqrt{P_{s2}} h_2 x_2 + n_{r,1}) + n_{d,i},$$

where $n_{d,i} \sim \mathscr{CN}(0, N_{di})$ is the noise sample at node i. Assuming perfect knowledge of $\boldsymbol{h}$ and b at S_i, the self-interference term created by its own transmitted signal x_i can be removed as $y_i = y_i' - \sqrt{P_{si} P_r} b h_i^2 x_i$. The received signal at S_i after removing the self-interference can then be written as

$$y_i = \sqrt{P_{sk} P_r} b h_1 h_2 x_k + \sqrt{P_r} b h_i n_r + n_{d,i} = \sqrt{P_{sk} P_r} b h_1 h_2 x_k + n_i, \tag{2.12}$$

where $k = 3 - i$ and the equivalent noise $n_i \sim \mathscr{CN}(0, N_{di} + \alpha_i b^2 P_r N_r)$.

Let $q_i = \mathbb{E}[|x_i|^2]$ so that the average power allocated at S_i in the first phase is $q_i P_{si}$. As before, two types of amplification coefficient can be used at relay depending on the availability of channel knowledge:

$$b = \sqrt{\frac{z_2}{q_1 P_{s1} \Xi_1 + q_2 P_{s2} \Xi_2 + N_r}}. \tag{2.13}$$

In the case of the FG CDI technique $\Xi = \phi$, whereas $\Xi = \alpha$ for the VG CI coefficient.

The conditional achievable rate in the $S_k \to S_i$ direction can be expressed from (2.12) as in [6, 8]:

$$I_i|\boldsymbol{h} = \frac{1}{2}\log\left(1 + \frac{q_k \alpha_1 \alpha_2 b^2 P_{sk} P_r}{N_{di} + \alpha_i b^2 P_r N_r}\right). \tag{2.14}$$

The conditional sum rate is then given as $I_{\text{sum}}|\boldsymbol{h} = I_1|\boldsymbol{h} + I_2|\boldsymbol{h}$ and the unconditional achievable sum rate is $I_{\text{sum}} = \mathbb{E}[I_{\text{sum}}|\boldsymbol{h}]$. It should be noted that the OW schemes introduced above can also be seen as TW protocols carried over four transmission phases, i.e., $S_1 \to S_2$ over the first two phases and $S_2 \to S_1$ over the remaining two.

2.3 FD Protocols

We now introduce the FD protocols of interest.

2.3.1 LR Protocol

First, we describe the *linear relaying* (LR) protocol introduced in [4], which is obtained by combining cooperative relaying in Fig. 1.2 with AF transmission. The LR protocol can be considered as a generalization of NAF relaying to FD operation and has only been studied in terms of achievable rate under no self-interference [2, 4]. As before, we assume that S_1 is the source S and S_2 the destination D. The transmission in LR is carried in frames composed of L symbol periods (where $T_c \geq L$ is a multiple of L). In each symbol period i $(1 \leq i \leq L)$, S transmits the information signal x_i to both the relay R and the destination D. At the same time, R transmits a linear combination of symbols previously received in the same frame to D.

Specifically, the signal received at the relay at time i can be expressed as

$$y_{r,i} = \sqrt{P_s} h_1 x_i + n_{r,i} + v_i, \tag{2.15}$$

where $n_{r,i} \sim \mathcal{CN}(0, N_r)$; and v_i is the residual self-interference term due to FD operation and after self-interference cancellation. In the following, we assume that this term is Gaussian as $v_i \sim \mathcal{CN}(0, V)$. Detailed discussions about the distribution and variance of v_i are provided in Chap. 7. The signal transmitted from the relay at time $i \geq 2$ can be written as

$$t_i = \sum_{k=1}^{i-1} b_{i,k} y_{r,k},$$

where $b_{i,k}$ is the coefficient used at time i to amplify the signal received at time k, $y_{r,k}$, $1 \leq k \leq i-1$. The signal received at the destination is thus given by

$$y_i = \sqrt{P_s} h_0 x_i + \sqrt{P_r} h_2 t_i + n_{d,i},$$

where $n_{d,i} \sim \mathcal{CN}(0, N_d)$. It should be noted that the relay operates in HD mode in the first time slot, i.e., it receives $y_{r,1}$ but does not transmit ($t_1 = 0$). As such, $y_{r,1}$ in (2.15) does not have the extra residual self-interference term and $v_1 = 0$.

The amplification coefficients $b_{i,k}$ can be grouped into a strictly lower triangular $L \times L$ amplification matrix $\boldsymbol{B}$ as

$$\boldsymbol{B} = \begin{cases} b_{i,k}, & 2 \leq i \leq L,\ 1 \leq k \leq i-1 \\ 0, & \text{o.w.} \end{cases}$$

The signal transmitted by the relay, $\boldsymbol{t} = [0, t_2, \ldots, t_L]^\top$, can then be written in vector form as

$$\boldsymbol{t} = \boldsymbol{B} \boldsymbol{y}_r = \boldsymbol{B}(\sqrt{P_s} h_1 \boldsymbol{x} + \boldsymbol{n}_r + \boldsymbol{v}), \tag{2.16}$$

where $\boldsymbol{y}_r = [y_{r,1}, \ldots, y_{r,L}]^\top$ is the vector received at the relay; $\boldsymbol{x} = [x_1, \ldots, x_L]^\top$; $\boldsymbol{n}_r = [n_{r,1}, \ldots, n_{r,L}]^\top \sim \mathcal{CN}(\boldsymbol{0}, N_r \boldsymbol{I}_L)$; and $\boldsymbol{v} = [0, v_2, \ldots, v_L]^\top \sim \mathcal{CN}(\boldsymbol{0}, \boldsymbol{K}_v)$ is the self-interference vector due to FD operation with

$$\boldsymbol{K}_v = V \cdot \begin{pmatrix} 0 & \boldsymbol{0} \\ \boldsymbol{0} & \boldsymbol{I}_{L-1} \end{pmatrix}. \tag{2.17}$$

The signal received at the destination over the L time slots can be similarly expressed in matrix form as

$$\boldsymbol{y} = \sqrt{P_s} \underbrace{(h_0 \boldsymbol{I}_L + \sqrt{P_r} h_1 h_2 \boldsymbol{B})}_{\boldsymbol{H}_{\text{LR}}} \boldsymbol{x} + \underbrace{\sqrt{P_r} h_2 \boldsymbol{B}(\boldsymbol{n}_r + \boldsymbol{v}) + \boldsymbol{n}_d}_{\boldsymbol{n}} = \sqrt{P_s} \boldsymbol{H}_{\text{LR}} \boldsymbol{x} + \boldsymbol{n}, \tag{2.18}$$

where $\boldsymbol{n}_d = [n_{d,1}, \ldots, n_{d,L}]^{\mathrm{T}} \sim \mathscr{CN}(\boldsymbol{0}, N_d \boldsymbol{I}_L)$. The equivalent $L \times L$ channel matrix in (2.18) is then given by

$$\boldsymbol{H}_{\mathrm{LR}} = h_0 \boldsymbol{I}_L + \sqrt{P_r} h_1 h_2 \boldsymbol{B}, \tag{2.19}$$

and the equivalent $L \times 1$ noise vector $\boldsymbol{n} \sim \mathscr{CN}(\boldsymbol{0}, \boldsymbol{K})$ with

$$\boldsymbol{K} = P_r \alpha_2 \boldsymbol{B}(N_r \boldsymbol{I}_L + \boldsymbol{K}_v)\boldsymbol{B}^{\dagger} + N_d \boldsymbol{I}_L. \tag{2.20}$$

Interestingly, the LR protocol in (2.18) reduces to the NAF in (2.1) for $L = 2$.

Let $\boldsymbol{Q} = \mathbb{E}[\boldsymbol{x}\boldsymbol{x}^{\dagger}]$ with diagonal elements q_i $(1 \le i \le L)$. The source then transmits an average power of $P_s \cdot \mathrm{tr}(\boldsymbol{Q})$ per transmission frame or $(P_s/L) \cdot \mathrm{tr}(\boldsymbol{Q})$ per symbol period. Similarly, let the covariance matrix of the signal transmitted by the relay be denoted as

$$\boldsymbol{Z} = \mathbb{E}[\boldsymbol{t}\boldsymbol{t}^{\dagger}] = \boldsymbol{B}(\alpha_1 P_s \boldsymbol{Q} + N_r \boldsymbol{I}_L + \boldsymbol{K}_v)\boldsymbol{B}^{\dagger}, \tag{2.21}$$

with diagonal elements z_i, $(1 \le i \le L)$. The relay then transmits an average power of $P_r \cdot \mathrm{tr}(\boldsymbol{Z})$ per transmission frame or $(P_r/L) \cdot \mathrm{tr}(\boldsymbol{Z})$ per symbol period.

Depending on the CSI available at the relay and on how the symbols $y_{r,k}$ are superimposed, different amplification matrices $\boldsymbol{B}$ can be considered as discussed before. For instance, one can assume that the relay only amplifies the signal received in the previous period similar to [2]. In this case, $b_{i,k} = b$ when $k = i - 1$ and $b_{i,k} = 0$ otherwise $(2 \le i \le L)$, where

$$b = \sqrt{\frac{\mathrm{tr}(\boldsymbol{Z})}{N_r(L-1) + V(L-2) + P_s \Xi_1 \sum_{i=2}^{L} q_{i-1}}}. \tag{2.22}$$

As before, $\Xi = \alpha$ when the relay has instantaneous knowledge of the S-R link and $\Xi = \phi$ when only second order statistics are known.

Assuming Gaussian codebooks at the source, the conditional achievable rate can be written from (2.18) as

$$I|\boldsymbol{h} = \frac{1}{L} \log\det(\boldsymbol{I}_L + \boldsymbol{H}_{\mathrm{LR}}{}^{\dagger} \boldsymbol{K}^{-1} \boldsymbol{H}_{\mathrm{LR}} \boldsymbol{Q}), \tag{2.23}$$

with $\boldsymbol{K}$ as in (2.20) and $\boldsymbol{H}_{\mathrm{LR}}$ as in (2.19).

2.3.2 DHAF Protocol

Next, we consider the FD DHAF protocol in Fig. 1.2. In this protocol, the direct S-D link is either non-existent $\alpha_0 = 0$ due to heavy shadowing or treated as a source of interference. In particular, the source continuously transmits its information symbol x_i while the relay amplifies and forwards the signal received in the previous time slot with a power amplification b. Specifically, the signal received at R at time i can be written as in (2.15), whereas the signal transmitted by R is now $t_i = by_{r,i-1}$ The signal received at D can then be written as

$$y_i = \sqrt{P_s P_r} h_1 h_2 b x_{i-1} + \sqrt{P_r} h_2 b (n_{r,i-1} + v_{i-1}) + \sqrt{P_s} h_0 x_i + n_{d,i}, \qquad (2.24)$$

where x_i is treated as interference when $\alpha_0 > 0$ and x_{i-1} is the desired symbol.

Let $\mathbb{E}[|x_i|^2] = q_1$ so that an average power of $q_1 P_s$ is spent at the source. Assuming an average power of $z_2 P_r$ at the relay, the amplification coefficient in (2.24) can be expressed as

$$b = \sqrt{\frac{z_2}{q_1 P_s \Xi_1 + N_r + V}}. \qquad (2.25)$$

Assuming Gaussian codebooks at S, the conditional achievable rate can now be written from (2.24) as

$$I|\boldsymbol{h} = \left(1 + \frac{\alpha_1 \alpha_2 q_1 P_s P_r b^2}{\alpha_2 P_r b^2 [N_r + V] + N_d + \alpha_0 q_1 P_s}\right). \qquad (2.26)$$

It can be observed that similar to the LR protocol, the relay of the DH protocol operates in HD mode in the first time slot. However, assuming that the DH transmission is carried continuously over several time slots, the effect of this transient state is negligible and can thus be ignored in (2.24).

2.4 Concluding Remarks

This chapter introduced the HD and FD protocols of interest. The achievable rate and error performance of these protocols are analyzed in the following chapters.

References

1. Azarian, K., Gamal, H.E., Schniter, P.: On the achievable diversity-multiplexing tradeoff in half-duplex cooperative channels. IEEE Trans. Inf. Theory **51**(12), 4152–4172 (2005). Doi:10.1109/TIT.2005.858920
2. Del Coso, A., Ibars, C.: Achievable rates for the AWGN channel with multiple parallel relays. IEEE Trans. Wireless Commun. **8**(5), 2524–2534 (2009). Doi:10.1109/TWC.2009.080288
3. Ding, Y., Zhang, J.K., Wong, K.M.: Ergodic channel capacities for the amplify-and-forward half-duplex cooperative systems. IEEE Trans. Inf. Theory **55**(2), 713–730 (2009). Doi:10.1109/TIT.2008.2009822

4. El Gamal, A., Mohseni, M., Zahedi, S.: Bounds on capacity and minimum energy-per-bit for AWGN relay channels. IEEE Trans. Inf. Theory **52**(4), 1545–1561 (2006). Doi:10.1109/TIT.2006.871579
5. Kim, S.J., Devroye, N., Tarokh, V.: Bi-directional half-duplex protocols with multiple relays. arXiv (2010). URL http://arxiv.org/abs/0810.1268v2
6. Kim, S.J., Devroye, N., Mitran, P., Tarokh, V.: Achievable rate regions and performance comparison of half-duplex bi-directional relaying protocols. IEEE Trans. Inf. Theory **57**(10), 6405–6418 (2011). Doi:10.1109/TIT.2011.2165132
7. Nabar, R.U., Bölcskei, H., Kneubühler, F.W.: Fading relay channel: Performance limits and space-time signal design. IEEE J. Sel. Areas Commun. **22**(6), 1099–1109 (2004). Doi:10.1109/JSAC.2004.830922
8. Rankov, B., Wittneben, A.: Spectral efficient protocols for half-duplex fading relay channels. IEEE J. Sel. Areas Commun. **25**(2), 379–389 (2007). Doi:10.1109/JSAC.2007.070213

Chapter 3
Half-Duplex AF Relaying: Capacity and Power Allocation Over Static Channels

During the past few years, much attention has been paid to the capacity, together with optimal power strategies, of various OW and TW HD-AF protocols over static channel links. Extensive research has been devoted to dual- or multi-hop AF systems that do not take the direct source-destination link into account. For instance, power allocation strategies that maximize the conditional achievable rate have been addressed for different configurations in [7, 8, 12, 13, 17] and references within. In particular, optimal power sharing strategies between the source and the relay to maximize the achievable rate in (2.11) were derived in [12]. Significant research has also been devoted to power allocation strategies for orthogonal AF schemes that take the direct link into consideration [5, 12, 14, 20]. Specifically, the maximization of (2.9) over the powers at the source and relay has been carried in [12]. For TWAF relaying, power allocation strategies for static channels have been also investigated in [4, 10, 16, 18]. In particular, the optimal power sharing scheme that maximizes the conditional sum rate in (2.14) has been derived in [10].

On the other hand, the analysis and optimization of OW non-orthogonal networks are more challenging and have received less attention [1, 6, 15]. For instance, the capacity of a NAF channel was investigated in the fast fading scenario in [6]. Under the assumption of full CSI available only at the destination node, it was shown in [6] that the input distribution must be Gaussian with a diagonal covariance matrix to achieve the ergodic capacity. Due to the complexity of the problem, the diagonal elements of the covariance matrix can only be obtained numerically in [6]. In addition, it was shown in [6] that the OAF and DT schemes are optimal in low and high signal-to-noise ratio (SNR) regimes, respectively. However, the maximization of the conditional rate for the static NAF channel in (2.5) has not been addressed in the literature.

The capacity of the static multiple-input multiple-output (MIMO) system has been well established in [19], where it was shown that the optimal signaling method is to water-fill over the square of the singular values of the channel matrix.

L. Jiménez Rodríguez et al., *Amplify-and-Forward Relaying in Wireless Communications*, SpringerBriefs in Computer Science,
DOI 10.1007/978-3-319-17981-0_3

Although the maximization for the NAF channel in (2.5) resembles that in static MIMO systems, there are significant differences that make the relay problem more challenging. Specifically, observe from (2.5) that different from MIMO systems, the channel and noise covariance matrices are functions of the powers transmitted at the source and relay. Thus, the technique proposed in [19] for MIMO systems is not applicable to the static NAF channel. More importantly, the mutual information of the NAF system in (2.5) is no longer concave and conventional optimization methods cannot be applied. Due to these challenging issues, closed-form solutions on the capacity limit of the general static NAF channel remain unknown, to the best of our knowledge.

Motivated by the above observations, this chapter attempts to provide some new results on the fundamental limit of the static cooperative AF channel. In particular, this chapter analyzes the capacity and characterizes the optimal input covariance matrix at the source and power allocation scheme at the relay for a half-duplex single-relay NAF system with static channel gains. Both individual and global power constraints are considered. Specifically, by focusing on the per-node power constrained system, we first derive all locally optimal solutions of the mutual information. By comparing these solutions, it is then shown that the capacity is achieved by using either a DT scheme, a NAF beamforming (BF) protocol with a specific unit-rank covariance matrix, or a NAF protocol using a specific full-rank (FR) covariance matrix. For the system under the joint power constraint, it is shown following a similar approach that the capacity can be achieved by using either a DT or a NAF-BF scheme. Thus, different from the fading scenario in [6], the OAF protocol is strictly suboptimal under both constraints and a diagonal covariance matrix is not always optimal. By further analyzing the asymptotic behavior in high and low transmission powers, it is demonstrated that NAF relaying is advantageous when the relay has large power and the source does not. However, the NAF protocol can provide significant gains over DT in medium transmission power regions as illustrated by different network configurations.

3.1 Problem Formulation

In this chapter, we consider the static scenario similar to [19]. In this scenario, the coherence time T_c is assumed to be large enough so that the channel gains $\boldsymbol{h} = [h_0, h_1, h_2]$ can be considered as time-invariant constants for the entire duration of the transmitted codeword. For this static system, a good estimate of the channel gains can be obtained and hence it is assumed that S, R and D have full knowledge of $\boldsymbol{h}$. Since the relay has full CSI, the CI coefficient $b = \sqrt{z_2/[q_1\alpha_1 + N_r]}$ in (2.4) is adopted. The input-output relation of the considered NAF system was explained in Sect. 2.2 and is given by (2.1). It is important to note from Sect. 2.2 that the NAF protocol is general in that it includes not only the OAF scheme in (2.8), but also the DT scheme in (2.6). Specifically, OAF corresponds to $q_2 = 0$, while DT is obtained by setting $z_2 = 0$. When $q_1, q_2, z_2 > 0$, one has the NAF protocol.

It can be shown from (2.1) that for a given transmit covariance matrix $\boldsymbol{Q} = \begin{pmatrix} q_1 & q_{12} \\ q_{12}^{\dagger} & q_2 \end{pmatrix}$ and power at the relay z_2, a Gaussian codebook must be used at the source to maximize the mutual information between the input and output vectors of the NAF channel [6, 19]. The mutual information between the input and output of the NAF channel is then given by (2.5). The objective of this chapter is to maximize the mutual information in (2.5) under both individual and global power constraints. First, for the per-node constraint scenario, we assume that $\text{tr}(\boldsymbol{Q}) = q_1 + q_2 \le q_s$ and $z_2 \le z_r$. With these constraints, the source and relay have respectively an average power constraint of $q_s P_s$ and $z_r P_r$ per cooperative frame, or $q_s P_s/2$ and $z_r P_r/2$ per symbol period. The constants q_s and z_r might be set to two to have constraints of P_s and P_r per symbol period. The per-node power-constrained capacity of the static NAF system can then be calculated as

$$C_{\text{indiv}} = \max_{\boldsymbol{Q}\succeq\boldsymbol{0},\, z_2\ge 0} I|\boldsymbol{h} \quad \text{s.t.} \quad \text{tr}(\boldsymbol{Q}) \le q_s,\ z_2 \le z_r. \tag{3.1}$$

To provide a fair comparison among different protocols and similar to [5, 7–9, 12, 14, 20], we also consider the global power constraint scenario. In this scenario, we assume that $P_s = P_r = P_t$ and $\text{tr}(\boldsymbol{Q}) + z_2 = q_1 + q_2 + z_2 \le q_t$. The system is then allowed to spend an average power of up to $q_t P_t$ per cooperative frame or $q_t P_t/2$ per symbol period. The global power-constrained capacity of the static NAF system is then calculated as

$$C_{\text{joint}} = \max_{\boldsymbol{Q}\succeq\boldsymbol{0},\, z_2\ge 0} I|\boldsymbol{h} \quad \text{s.t.} \quad \text{tr}(\boldsymbol{Q}) + z_2 \le q_t. \tag{3.2}$$

Observe from (2.5) that, different from the static MIMO system in [19], the channel and noise covariance matrices are functions of the powers transmitted at the source and relay, i.e., $\boldsymbol{H}_{\text{NAF}} = \boldsymbol{H}_{\text{NAF}}(q_1, z_2)$ and $\boldsymbol{K} = \boldsymbol{K}(q_1, z_2)$. As a consequence, the problems in (3.1) and (3.2) are in general not concave with respect to z_2 and the psd matrix $\boldsymbol{Q}$. Hence, water-filling over the eigenvalues of $\boldsymbol{H}_{\text{NAF}}^{\dagger}\boldsymbol{K}^{-1}\boldsymbol{H}_{\text{NAF}}$ is no longer optimal and the Karush-Kuhn-Tucker (KKT) conditions cannot be used to find the global maximizer. In the following, we first derive all local maximizers of the mutual information in (2.5). The capacities in (3.1) and (3.2) can then be calculated by comparing these local solutions.

3.2 Capacity Under Individual Power Constraints

In this section, we derive the optimal covariance matrix at the source and optimal power allocation scheme at the relay to achieve the capacity with the per-node power constraints in (3.1). Let $q_{12} = qe^{j\theta}$. To simplify the notation, let also $\gamma_0 = (P_s\alpha_0)/N_d$, $\gamma_1 = (P_s\alpha_1)/N_r$ and $\gamma_2 = (P_r\alpha_2)/N_d$. In the remainder of the chapter, we assume that the values of γ_l are arbitrary but strictly positive.

Substituting the amplification coefficient $b = \sqrt{z_2/[q_1\alpha_1 + N_r]}$, the mutual information in (2.5) can be written as

$$I|\boldsymbol{h} = \frac{1}{2}\log\det(\boldsymbol{I}_2 + P_s\boldsymbol{H}^{\dagger}_{\text{NAF}}\boldsymbol{K}^{-1}\boldsymbol{H}_{\text{NAF}}\boldsymbol{Q}) = \frac{1}{2}\log\left[f_C(\boldsymbol{q}, q_{12})\right], \tag{3.3}$$

where $\boldsymbol{q} = [q_1, q_2, z_2]$ and

$$f_C(\boldsymbol{q}, q_{12}) = 1 + f_d(\boldsymbol{q}) + f_o(\boldsymbol{q}, q_{12}), \tag{3.4}$$

with

$$f_d(\boldsymbol{q}) = \frac{q_1\gamma_0 + q_2\gamma_0 + q_1^2\gamma_0\gamma_1 + q_1^2q_2\gamma_1\gamma_0^2 + q_1q_2\gamma_0^2 + q_1q_2\gamma_1\gamma_0 + q_1z_2\gamma_2\gamma_0 + q_1z_2\gamma_1\gamma_2}{q_1\gamma_1 + z_2\gamma_2 + 1} \tag{3.5}$$

and

$$f_o(\boldsymbol{q}, q_{12}) = \frac{2q\sqrt{z_2\gamma_0\gamma_1\gamma_2(q_1\gamma_1 + 1)}\cos(\theta_0 - \theta_1 - \theta_2 - \theta) - q^2\gamma_0^2(q_1\gamma_1 + 1)}{q_1\gamma_1 + z_2\gamma_2 + 1}. \tag{3.6}$$

Given that $\log(\cdot)$ is a monotonically increasing function, the capacity can be obtained by maximizing $f_C(\boldsymbol{q}, q_{12})$ in (3.4). For the system with per-node power constraints, the capacity in (3.1) can be obtained by solving the following optimization problem:

$$\max_{\boldsymbol{q}, q_{12}} \quad f_C(\boldsymbol{q}, q_{12}) \quad \text{s.t.} \quad \begin{cases} q, q_1, q_2, z_2 \geq 0, \\ q_1 + q_2 \leq q_s, \; z_2 \leq z_r, \\ q^2 \leq q_1 q_2, \end{cases} \tag{3.7}$$

where the last constraint guarantees that $\boldsymbol{Q}$ remains psd. The feasible region of (3.7) is illustrated in Fig. 3.1.

Fig. 3.1 Feasible region for individual power constraints

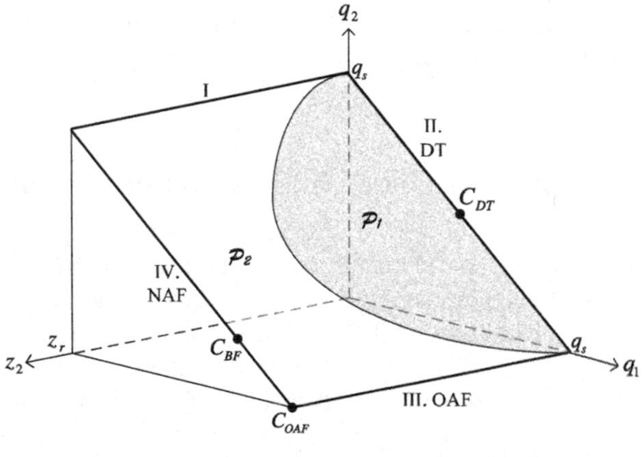

First, observe from (3.4) that $f_C(\cdot)$ is an increasing function of q_2. At the optimal solution, the source must then use all its power, i.e., $q_1 + q_2 = q_s$. Define the plane $\mathscr{R}_1 = \{\boldsymbol{q} \in \mathbb{R}^3 \mid q_k, z_2 \geq 0,\ z_2 \leq z_r,\ q_1 + q_2 = q_s\}$. The solution to (3.7) then lies on $\mathscr{R}_1$, i.e., on the front surface in Fig. 3.1. Furthermore, it can be seen from (3.6) that $f_o(\boldsymbol{q}, qe^{j\theta}) \leq f_o(\boldsymbol{q}, qe^{j[\theta_0-\theta_1-\theta_2]})$ for any feasible $\boldsymbol{q}$ and $q = |q_{12}|$. The equality is achieved when $\theta^\star = \theta_0 - \theta_1 - \theta_2$, which becomes the optimal angle. Substituting this optimal angle, $f_o(\cdot)$ can be written as $f_o(\boldsymbol{q}, qe^{j\theta^\star}) = Aq^2 + Bq$ with A and B as in (3.6). Since $A < 0$, $f_o(\boldsymbol{q}, qe^{j\theta^\star})$ is a concave function of q. Given that the maximum value of q is $\sqrt{q_1 q_2}$ due to the psd constraint in (3.7), the optimal q is either at the stationary point $q = -B/2A$ or at this maximum value, i.e.,

$$q^\star = \min\left\{\sqrt{\frac{z_2\gamma_1\gamma_2}{\gamma_0^3(q_1\gamma_1 + 1)}}, \sqrt{q_1 q_2}\right\}. \tag{3.8}$$

Then,

$$f_C(\boldsymbol{q}, qe^{j\theta}) \leq f_C(\boldsymbol{q}, q_{12}^\star) = 1 + f_d(\boldsymbol{q}) + f_o(\boldsymbol{q}, q_{12}^\star), \tag{3.9}$$

where the equality is achieved when $q_{12}^\star = q^\star e^{j\theta^\star}$. For any given $\boldsymbol{q}$, the optimal value of q_{12} is thus $q_{12}^\star$ with $\theta^\star = \theta_0 - \theta_1 - \theta_2$ and $q^\star$ as in (3.8).

Depending on whether $\sqrt{[z_2\gamma_1\gamma_2]/[\gamma_0^3(q_1\gamma_1 + 1)]}$ in (3.8) is smaller than $\sqrt{q_1 q_2}$, $\mathscr{R}_1$ can be divided into two regions. Let

$$P(q_1) = \frac{\gamma_0^3}{\gamma_1\gamma_2} q_1(q_s - q_1)(q_1\gamma_1 + 1), \tag{3.10}$$

be a cubic polynomial in q_1 obtained by setting $z_2\gamma_1\gamma_2 = q_1 q_2 \gamma_0^3(q_1\gamma_1 + 1)$. As shown in Fig. 3.1, define the regions

$$\mathscr{P}_1 = \{\boldsymbol{q} \in \mathscr{R}_1 \mid z_2 \leq P(q_1)\}, \quad \text{and} \quad \mathscr{P}_2 = \{\boldsymbol{q} \in \mathscr{R}_1 \mid z_2 \geq P(q_1)\}. \tag{3.11}$$

Let $f_{\mathscr{P}_1}(\boldsymbol{q}) = f_C(\boldsymbol{q}, \sqrt{[z_2\gamma_1\gamma_2]/[\gamma_0^3(q_1\gamma_1 + 1)]}e^{j\theta^\star})$ and $f_{\mathscr{P}_2}(\boldsymbol{q}) = f_C(\boldsymbol{q}, \sqrt{q_1 q_2}e^{j\theta^\star})$. Then,

$$f_C(\boldsymbol{q}, q_{12}^\star) = \begin{cases} f_{\mathscr{P}_1}(\boldsymbol{q}), & \boldsymbol{q} \in \mathscr{P}_1 \\ f_{\mathscr{P}_2}(\boldsymbol{q}), & \boldsymbol{q} \in \mathscr{P}_2. \end{cases} \tag{3.12}$$

We have the following lemma regarding $f_C(\boldsymbol{q}, q_{12}^\star)$.

Lemma 3.1. *$f_C(\boldsymbol{q}, q_{12}^{\star})$ is not quasiconcave in $\mathscr{R}_1$ and has no local maximizers in the interior of $\mathscr{R}_1$.*

Proof. This follows from the fact that the only stationary point of $f_C(\boldsymbol{q}, q_{12}^{\star})$ in the interior of $\mathscr{R}_1$ is a saddle point (please refer to [11]). □

From Lemma 3.1, neither $f_C(\boldsymbol{q}, q_{12}^{\star})$ nor consequently $f_C(\boldsymbol{q}, q_{12})$ in (3.4) are quasiconcave in $\mathscr{R}_1$ [3, Chap. 3.4]. Thus, $f_C(\boldsymbol{q}, q_{12})$ cannot be log-concave [3]. As such, (3.1) and (3.7) are non-concave optimization problems. Given that $f_C(\boldsymbol{q}, q_{12}^{\star})$ does not have any local maximizers in the interior of $\mathscr{R}_1$, the solution to (3.7) must lie on the boundary. As illustrated in Fig. 3.1, let the line segments that form the perimeter of the rectangle $\mathscr{R}_1$ be defined as $\ell_{\mathrm{I}} = \{\boldsymbol{q} \in \mathbb{R}^3 \mid q_1 = 0,\ q_2 = q_s,\ 0 \le z_2 \le z_r\}$, $\ell_{\mathrm{II}} = \{\boldsymbol{q} \in \mathbb{R}^3 \mid z_2 = 0,\ q_1 + q_2 = q_s,\ q_1, q_2 \ge 0\}$, $\ell_{\mathrm{III}} = \{\boldsymbol{q} \in \mathbb{R}^3 \mid q_2 = 0,\ q_1 = q_s,\ 0 \le z_2 \le z_r\}$, and $\ell_{\mathrm{IV}} = \{\boldsymbol{q} \in \mathbb{R}^3 \mid z_2 = z_r,\ q_1 + q_2 = q_s,\ q_1, q_2 \ge 0\}$. Note that ℓ_{II}, ℓ_{III} and ℓ_{IV} correspond to the DT, OAF, and NAF protocols, respectively, while ℓ_{I} is the trivial case in which the source transmits only in the second time slot and the relay amplifies noise. We first analyze the local solutions in ℓ_{I}, ℓ_{II} and ℓ_{III}.

Lemma 3.2. *The maximizer in ℓ_{II}, denoted as C_{DT}, is achieved by the DT scheme with equal power allocation:*

$$q_{1,DT} = q_{2,DT} = q_s/2, \quad z_{2,DT} = q_{12,DT} = 0, \tag{3.13}$$

and outperforms the maximizer in ℓ_I. The maximizer in ℓ_{III}, denoted as C_{OAF}, is achieved by the OAF scheme with full power at both nodes:

$$q_{1,OAF} = q_s, \quad z_{2,OAF} = z_r, \quad q_{2,OAF} = q_{12,OAF} = 0. \tag{3.14}$$

Proof. When $\boldsymbol{q} \in \ell_{\mathrm{I}} \subset \mathscr{P}_2$, $q_1 = 0$ and $q^{\star} = 0$ from (3.8). Since $f_{\mathscr{P}_2}([0, q_s, z_2])$ is a decreasing function of z_2, it is maximized when $z_2 = 0$. When $\boldsymbol{q} \in \ell_{\mathrm{II}} \subset \mathscr{P}_1$, $z_2 = 0$ and $q^{\star} = 0$ from (3.8). The function $f_{\mathscr{P}_1}([q_1, q_2, 0]) = (q_1\gamma_0 + 1)(q_2\gamma_0 + 1)$ is maximized when $q_1 = q_2 = q_s/2$ from the geometric/arithmetic mean inequality. Furthermore, $f_C([q_s/2, q_s/2, 0], 0) > f_C([0, q_s, 0], 0)$ for any $\gamma_0 > 0$. Thus, the maximizer in ℓ_{II} is greater than the one in ℓ_{I}. When $\boldsymbol{q} \in \ell_{\mathrm{III}} \subset \mathscr{P}_2$, $q_2 = 0$ and $q^{\star} = 0$ from (3.8). Since $f_{\mathscr{P}_2}([q_s, 0, z_2])$ is increasing with z_2, it is maximized when $z_2 = z_r$. □

We now analyze the maximizer in the remaining segment ℓ_{IV}.

Lemma 3.3. *The maximizer in ℓ_{IV}, denoted as C_{NAF}, is given by*

$$C_{NAF} = \max\{C_{BF}, C_{FR}\}, \tag{3.15}$$

and outperforms the maximizer in ℓ_{III} since $C_{BF} > C_{OAF}$. In (3.15), C_{BF} is achieved by

$$q_{1,BF} = r_6, \quad q_{2,BF} = q_s - r_6, \quad z_{2,BF} = z_r, \quad q_{12,BF} = \sqrt{q_{1,BF}, q_{2,BF}} e^{j\theta^\star}, \tag{3.16}$$

where r_6 is the only root in $0 < r_5 < r_6 < q_s$ of the 6th order polynomial

$$P_6(q_1) = \gamma_0\gamma_1 P_5(q_1)^2 - z_r\gamma_2 A_8^2 q_1(q_s - q_1)(q_1\gamma_1 + 1), \tag{3.17}$$

with $A_8 = -\gamma_1(3z_r^2\gamma_2^2 + 4z_r\gamma_2 + q_s\gamma_1 + 2z_r q_s\gamma_1\gamma_2 + 1)$, and r_5 is the unique positive root of the cubic polynomial

$$P_5(q_1) = A_5 q_1^3 + B_5 q_1^2 + C_5 q_1 + D_5, \tag{3.18}$$

with $A_5 = -\gamma_1^2$, $B_5 = -3\gamma_1(z_r\gamma_2 + 1)$, $C_5 = q_s\gamma_1 - 2z_r\gamma_2 + 2z_r q_s\gamma_1\gamma_2 - 2$, and $D_5 = q_s(z_r\gamma_2 + 1)$; whereas C_{FR} is achieved by

$$q_{1,FR} = r_3, \quad q_{2,FR} = q_s - r_3, \quad z_{2,FR} = z_r, \quad q_{12,FR} = \sqrt{\frac{z_{2,FR}\gamma_1\gamma_2}{\gamma_0^3(q_{1,FR}\gamma_1 + 1)}} e^{j\theta^\star}, \tag{3.19}$$

as long as $P(r_3) - z_r > 0$, where $P(q_1)$ is the cubic polynomial in (3.10) and r_3 is the largest real root of the cubic polynomial

$$P_3(q_1) = A_3 q_1^3 + B_3 q_1^2 + C_3 q_1 + D_3, \tag{3.20}$$

with $A_3 = -2\gamma_0^3\gamma_1^2$, $B_3 = \gamma_0^3\gamma_1(q_s\gamma_1 - 3z_r\gamma_2 - 4)$, $C_3 = 2\gamma_0^3(q_s\gamma_1 - 1)(z_r\gamma_2 + 1)$, and $D_3 = z_r^2\gamma_0^2\gamma_2^2 + z_r^2\gamma_0\gamma_1\gamma_2^2 + q_s z_r\gamma_0^3\gamma_2 + q_s z_r\gamma_0^2\gamma_1\gamma_2 + z_r\gamma_0^2\gamma_2 + z_r\gamma_0\gamma_1\gamma_2 - z_r\gamma_1^2\gamma_2 + q_s\gamma_0^3$.

Proof. When $\boldsymbol{q} \in \ell_{\rm IV}$, we must consider two subcases depending on whether $P(q_1)$ is smaller or greater than z_r for $0 \le q_1 \le q_s$. Denote the line subsegments $\ell_{\rm IV}^{\mathscr{P}_1}(z_r) = \{\boldsymbol{q} \in \ell_{\rm IV} \cap \mathscr{P}_1\} = \{\boldsymbol{q} \in \ell_{\rm IV} \mid z_r \le P(q_1)\}$ and $\ell_{\rm IV}^{\mathscr{P}_2}(z_r) = \{\boldsymbol{q} \in \ell_{\rm IV} \cap \mathscr{P}_2\} = \{\boldsymbol{q} \in \ell_{\rm IV} \mid z_r \ge P(q_1)\}$. When $P(q_1) < z_r$, $\ell_{\rm IV}$ is completely in $\mathscr{P}_2$, i.e., $\ell_{\rm IV} = \ell_{\rm IV}^{\mathscr{P}_2}(z_r)$ and $\ell_{\rm IV}^{\mathscr{P}_1}(z_r) = \emptyset$. On the other hand, when $P(q_1) \ge z_r$, the two extremes of $\ell_{\rm IV}$ are in $\mathscr{P}_2$ while the mid-section is in $\mathscr{P}_1$. The maximizers for these two subcases are derived in [11]. □

First, observe from Lemma 3.3 that $\ell_{\rm IV}$ has at most two maximizers. Analogous to MIMO systems [2, Chap. 2.3], we have denoted the solution in (3.16) as the NAF beamforming (NAF-BF) scheme. This is because the rank of $\boldsymbol{Q}$ is unity when $|q_{12,\rm BF}| = \sqrt{q_{1,\rm BF}, q_{2,\rm BF}}$ ($\det(\boldsymbol{Q}) = 0$), which corresponds to the case where the source transmits the same information symbol in both time slots. On the other hand, the solution in (3.19) corresponds to the NAF protocol with a full-rank covariance matrix, NAF-FR. Although the NAF-BF maximizer in (3.16) is always feasible, note from Lemma 3.3 that the NAF-FR one in (3.19) might not depending on the channel gains, i.e., when $P(r_3) - z_r < 0$. More importantly, it can be implied from Lemma 3.3 that the maximizer in ℓ_{III} cannot be the solution to (3.7), i.e., the OAF protocol is not optimal under individual constraints. This is because the NAF-BF maximizer is always feasible and outperforms the OAF one. Finally, it is also important to note that different from static MIMO systems [19], the NAF solutions in (3.16) and (3.19) do not diagonalize $\boldsymbol{H}_{\text{NAF}}^{\dagger}\boldsymbol{K}^{-1}\boldsymbol{H}_{\text{NAF}}$ in (3.3).

From the above lemmas, the maximizer of (3.7) and hence the capacity in (3.1) is finally given in the following theorem.

Theorem 3.1. *The capacity of the system under individual power constraints C_{indiv} in (3.1) is given by*

$$C_{indiv} = \max\{C_{DT}, C_{BF}, C_{FR}\}. \tag{3.21}$$

Proof. This follows directly from Lemmas 3.1–3.3. □

Although the capacity in Theorem 3.1 is expressed as the maximum rate among these three schemes, the solution in (3.21) provides several important insights. For instance, among all possible transmission protocols, the capacity in (3.21) can only be achieved by the DT scheme with equal power allocation as in (3.13), the NAF-BF scheme with the power allocation in (3.16), or the NAF-FR scheme with a full-rank covariance matrix as in (3.19). It can also be seen from (3.21) that at the optimal solution, the relay uses either all its available power or no power at all. This is in contrast to the behavior of AF systems over fading channels in [6], where it was shown that the relay power can take on any value between zero and z_r. However, similar to [6], using full power at the relay is in fact not beneficial under some channel conditions and DT is optimal. Finally, note from (3.21) that different from the fading channel in [6], a diagonal covariance matrix is only optimal for DT.

3.3 Capacity Under Joint Power Constraints

In the previous section, we consider the system under individual power constraints. We now analyze the capacity under a joint power constraint. For the system with a global power constraint, the capacity in (3.2) can be obtained by solving the following optimization problem:

$$\max_{\boldsymbol{q},q_{12}} \quad f_C(\boldsymbol{q},q_{12}) \quad \text{s.t.} \quad \begin{cases} q, q_1, q_2, z_2 \geq 0, \\ q_1 + q_2 + z_2 \leq q_t, \\ q^2 \leq q_1 q_2, \end{cases} \tag{3.22}$$

with $f_C(\boldsymbol{q},q_{12})$ as in (3.4). The feasible region of (3.22) is illustrated in Fig. 3.2.

First, at the optimal solution, it is easy to show that the system must use all its power $q_1+q_2+z_2=q_t$. This follows again from the fact that $f_C(\cdot)$ is an increasing function of q_2. The solution to (3.7) then must lie on the front plane in Fig. 3.2, i.e.,

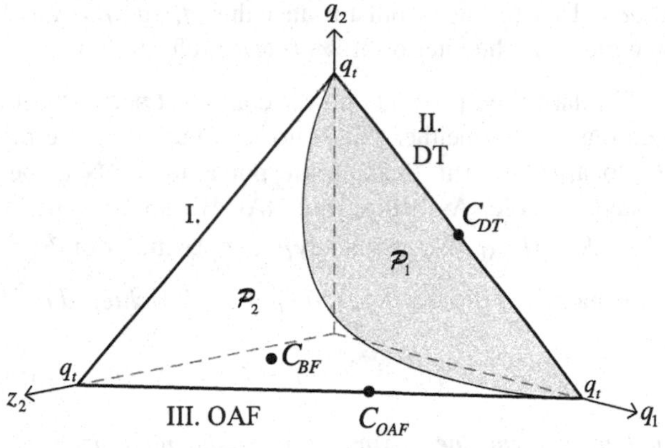

Fig. 3.2 Feasible region for joint power constraint

on the plane $\mathscr{R}_2 = \{\boldsymbol{q} \in \mathbb{R}^3 \mid 0 \leq q_k, z_2 \leq q_t,\ q_1 + q_2 + z_2 = q_t\}$. Define the line segments that form the perimeter of the triangle $\mathscr{R}_2$ as $\ell_{\mathrm{I}} = \{\boldsymbol{q} \in \mathbb{R}^3 \mid q_1 = 0,\ q_2 + z_2 = q_t,\ q_2, z_2 \geq 0\}$, $\ell_{\mathrm{II}} = \{\boldsymbol{q} \in \mathbb{R}^3 \mid z_2 = 0,\ q_1 + q_2 = q_t,\ q_1, q_2 \geq 0\}$, and $\ell_{\mathrm{III}} = \{\boldsymbol{q} \in \mathbb{R}^3 \mid q_2 = 0,\ q_1 + z_2 = q_t,\ q_1, z_2 \geq 0\}$. As before, ℓ_{II}, ℓ_{III}, and ℓ_{I} represent the DT, OAF and trivial protocols, respectively, while the interior of $\mathscr{R}_2$ now corresponds to the NAF scheme.

Similar to (3.9) in the previous section, $f_C(\boldsymbol{q}, q_{12}) \leq f_C(\boldsymbol{q}, q_{12}^\star)$ for any feasible $\boldsymbol{q}$, where the equality is achieved when $q_{12}^\star = q^\star e^{j\theta^\star}$ with $q^\star$ as in (3.8) and $\theta^\star = \theta_0 - \theta_1 - \theta_2$. As shown in Fig. 3.2, $\mathscr{R}_2$ can then be divided into two regions as

$$\mathscr{P}_1 = \{\boldsymbol{q} \in \mathscr{R}_2 \mid q_2 \geq P(q_1)\}, \quad \text{and} \quad \mathscr{P}_2 = \{\boldsymbol{q} \in \mathscr{R}_2 \mid q_2 \leq P(q_1)\}, \tag{3.23}$$

where now

$$P(q_1) = \frac{\gamma_1\gamma_2(q_t - q_1)}{\gamma_0^3 q_1(q_1\gamma_1 + 1) + \gamma_1\gamma_2}. \tag{3.24}$$

Hence, $f_C(\boldsymbol{q}, q_{12}^\star)$ is again given as in (3.12) but now using the regions defined by (3.23) and (3.24). We first analyze $f_C(\boldsymbol{q}, q_{12}^\star)$ the interior of $\mathscr{R}_2$.

Lemma 3.4. *$f_C(\boldsymbol{q}, q_{12}^\star)$ is not quasiconcave in $\mathscr{R}_2$ and has a single local maximizer in the interior of $\mathscr{R}_2$, denoted as C_{BF}, that is achieved by*

$$q_{1,BF} = \begin{cases} q_t/2, & \gamma_1 = \gamma_2 \\ \frac{-(q_t\gamma_2+1)+\sqrt{(q_t\gamma_1+1)(q_t\gamma_2+1)}}{\gamma_1-\gamma_2}, & \gamma_1 \neq \gamma_2, \end{cases}$$

$$q_{2,BF}(q_{1,BF}) = \frac{\gamma_0(q_t - q_{1,BF})[q_{1,BF}\gamma_1 + (q_t - q_{1,BF})\gamma_2 + 1]^2}{\gamma_0[q_{1,BF}\gamma_1 + (q_t - q_{1,BF})\gamma_2 + 1]^2 + q_{1,BF}\gamma_1\gamma_2(q_{1,BF}\gamma_1 + 1)},$$

$$z_{2,BF} = q_t - q_{1,BF} - q_{2,BF}, \quad q_{12,BF} = \sqrt{q_{1,BF}, q_{2,BF}}e^{j\theta^\star}. \tag{3.25}$$

Proof. This follows from the fact that $f_C(\boldsymbol{q}, q_{12}^\star)$ has both a saddle and a local maximizer in the interior of $\mathscr{R}_2$ (please refer to [11]). □

Similar to the per-node power constraint scenario in Lemma 3.1, observe from Lemma 3.4 that neither (3.22) nor (3.2) are concave problems [3]. Note also that the local solution in (3.25) corresponds to the NAF beamforming scheme and is always feasible. As before, the NAF-BF solution in (3.25) does not diagonalize $\boldsymbol{H}_{\mathrm{NAF}}^\dagger \boldsymbol{K}^{-1} \boldsymbol{H}_{\mathrm{NAF}}$. We now analyze the perimeter of $\mathscr{R}_2$.

Lemma 3.5. *The maximizer in ℓ_{II}, C_{DT}, is achieved by*

$$q_{1,DT} = q_{2,DT} = q_t/2, \quad z_{2,DT} = q_{12,DT} = 0, \tag{3.26}$$

and outperforms the maximizer in ℓ_I. The maximizer in ℓ_{III}, C_{OAF}, is achieved by

$$q_{1,OAF} = \begin{cases} q_t, & b_1 \geq 0 \\ \min\{q_t, -c_1/b_1\}, & \gamma_1 = \gamma_2 \\ \min\left\{q_t, \frac{-b_1 - \sqrt{b_1^2 - 4a_1c_1}}{2a_1}\right\}, & b_1 < 0, \gamma_1 \neq \gamma_2 \end{cases}$$

$$z_{2,OAF} = q_t - q_{1,OAF}, \quad q_{2,OAF} = q_{12,OAF} = 0, \tag{3.27}$$

where $a_1 = (\gamma_1 - \gamma_2)(\gamma_0\gamma_1 - \gamma_0\gamma_2 - \gamma_1\gamma_2)$, $b_1 = 2(q_t\gamma_2 + 1)(\gamma_0\gamma_1 - \gamma_0\gamma_2 - \gamma_1\gamma_2)$ and $c_1 = (q_t\gamma_2 + 1)(\gamma_0 + q_t\gamma_0\gamma_2 + q_t\gamma_1\gamma_2)$.

Proof. The proof for ℓ_{I} and ℓ_{II} follows from Lemma 3.2 by replacing q_s by q_t. The proof for ℓ_{III} is given in [12, 20]. □

The maximizer of (3.22) and thus the capacity in (3.2) are finally given in the following theorem.

Theorem 3.2. *The capacity of the system under joint power constraints C_{joint} in (3.2) is given by*

$$C_{joint} = \max\{C_{DT}, C_{BF}\}. \tag{3.28}$$

Proof. This follows from Lemmas 3.4–3.5, and the fact that $C_{\mathrm{BF}} > C_{\mathrm{OAF}}$ [11]. □

The capacity in (3.28) is then achieved by either the DT scheme with the power allocation in (3.26), or the NAF-BF scheme with the power allocation in (3.25). Different from Theorem 3.1, observe from (3.28) that the capacity cannot be achieved by a full-rank covariance matrix. However, similar to the individual power constraint scenario, note from Theorem 3.2 that the OAF protocol is strictly suboptimal since $C_{\mathrm{BF}} > C_{\mathrm{OAF}}$. As such, orthogonal transmission is not optimal under both individual and joint power constraints. This is directly in contrast to the AF system over fading channels in [6], where it was shown that the OAF scheme is optimal in low SNR regimes.

3.4 High and Low Power Regions

In the previous section, the capacity of the static AF system was derived and expressions for the optimal covariance matrix at the source and power allocation at the relay were provided. In this section, we provide further insights on the behavior of the capacities in (3.21) and (3.28) along with their respective allocations by focusing on two asymptotic regions, namely the low and high transmission powers. Although the OAF protocol has been shown to be suboptimal, its capacity will also be discussed here due to its popularity in the literature.

3.4.1 *Per-Node Power Constraints*

As shown in Theorem 3.1, the capacity for the individual power constraint scenario is achieved by either the DT, the NAF-FR or the NAF-BF protocol. Hence, one must evaluate and compare C_{DT}, C_{BF} and C_{FR} in (3.21). Consider the following asymptotic cases.

3.4.1.1 High Source Power, Fixed Relay Power

In this case, we assume that $P_s \to \infty$ while P_r remains fixed. Substituting (3.13) in (3.3), the capacity of the DT scheme is given as $C_{\mathrm{DT}} = \log(1 + [q_s\gamma_0]/2) = \log(P_s) + O(1)$, which grows logarithmically with P_s as already well-known. For the NAF-FR scheme, it can be easily shown that $r_3 \to q_s/2$. Given that $P(q_s/2) - z_r > 0$ for large P_s, $q_{1,\mathrm{FR}} \to q_s/2$ from Lemma 3.3. Substituting (3.19) in (3.4) with $q_{1,\mathrm{FR}} = q_s/2$, $f_C(\boldsymbol{q}_{\mathrm{FR}}, q_{12,\mathrm{FR}}) \approx ([q_s\gamma_0]/2)^2$ and the capacity of the NAF-FR scheme $C_{\mathrm{FR}} \approx \log([q_s\gamma_0]/2) \approx \log(P_s)$ also grows logarithmically with P_s. However, by comparing $(1+[q_s\gamma_0]/2)^2$ to (3.4) with $q_{1,\mathrm{FR}} = q_s/2$ in (3.19), $C_{\mathrm{DT}} > C_{\mathrm{FR}}$ for sufficiently large P_s. The NAF-FR system then approaches the DT scheme from below. Moreover, it is easy to see that $C_{\mathrm{BF}} \approx [1/2]\log(P_s)$, i.e., the capacity of the NAF-BF scheme only increases logarithmically with $\sqrt{P_s}$. This is due to the fact that a single information symbol is sent in two time slots. Due to the same reason, $C_{\mathrm{OAF}} \approx [1/2]\log(P_s)$. Since $C_{\mathrm{BF}} > C_{\mathrm{OAF}}$ from Lemma 3.3, the OAF scheme approaches the NAF-BF one from below. Therefore, for large source powers, $C_{\mathrm{DT}} > C_{\mathrm{FR}} > C_{\mathrm{BF}} > C_{\mathrm{OAF}}$ with $C_{\mathrm{FR}} \to C_{\mathrm{DT}}$ and $C_{\mathrm{OAF}} \to C_{\mathrm{BF}}$. The capacity in (3.21) is then achieved by the DT scheme.

3.4.1.2 High Relay Power, Fixed Source Power

In this case, we assume that $P_r \to \infty$ and P_s remains fixed. When P_r is sufficiently large, it is easy to see that $P(q_1) - z_r < 0$ in the entire range $0 \leq q_1 \leq q_s$. The NAF-FR maximizer in (3.19) is then not feasible from Lemma 3.3. For the

NAF-BF scheme, $q_{1,\mathrm{BF}} = r_6 \to q_s$. From (3.16), (3.14) and the fact that $C_{\mathrm{BF}} > C_{\mathrm{OAF}}$, this means that the capacity of the OAF scheme approaches that of the NAF-BF protocol from below, i.e., $C_{\mathrm{OAF}} \to C_{\mathrm{BF}}$. Substituting (3.14) in (3.3), the capacity of the OAF system can then be approximated as $C_{\mathrm{OAF}} \approx [1/2]\log(1 + q_s[\gamma_0 + \gamma_1])$, which outperforms $C_{\mathrm{DT}} = \log(1 + [q_s\gamma_0]/2)$ as long as $4\gamma_1 > q_s\gamma_0^2$. Hence, for large relay powers, $C_{\mathrm{OAF}} \to C_{\mathrm{BF}}$ and $C_{\mathrm{BF}} > C_{\mathrm{OAF}} > C_{\mathrm{DT}}$ when $q_s\gamma_0^2 < 4\gamma_1$, or $C_{\mathrm{DT}} > C_{\mathrm{BF}} > C_{\mathrm{OAF}}$ when $q_s\gamma_0^2 > 4\gamma_1$. The capacity in (3.21) is then achieved by either the NAF-BF or the DT scheme depending on the values of q_s, γ_0 and γ_1.

3.4.1.3 High Source and Relay Power

We now consider the scenario in which both $P_s, P_r \to \infty$ but their ratio P_r/P_s remains constant. As in first case, it can be easily shown that $C_{\mathrm{FR}} \approx \log(P_s)$ and $C_{\mathrm{BF}} \approx [1/2]\log(P_s)$, $C_{\mathrm{DT}} \approx \log(P_s)$ and $C_{\mathrm{OAF}} \approx [1/2]\log(P_s)$. By comparing the function values in (3.4) with (3.19) and (3.13), it can further be shown that $C_{\mathrm{DT}} > C_{\mathrm{FR}}$. Therefore, $C_{\mathrm{DT}} > C_{\mathrm{FR}} > C_{\mathrm{BF}} > C_{\mathrm{OAF}}$. Thus, as in the first case, DT is optimal when both the source and relay have high powers.

3.4.1.4 Low Source and Relay Power

We finally consider the scenario in which $P_s, P_r \to 0$ with P_r/P_s fixed. For the DT scheme, $C_{\mathrm{DT}} = \log(1 + [q_s\gamma_0]/2) \approx [q_s\gamma_0]/[2\ln(2)]$, where we have used the fact that $\ln(1 + x) \approx x$ for small values of x. The capacity of the DT scheme then decreases linearly with P_s as already-known. For the NAF-FR protocol, the maximizer in (3.19) is not feasible as $P(q_1) - z_r < 0$ when $P_s, P_r \to 0$. For the NAF-BF scheme, $q_{1,\mathrm{BF}} = r_6 \to q_s/2$. Substituting (3.16) in (3.4) with $q_{1,\mathrm{BF}} = q_s/2$, $C_{\mathrm{BF}} \approx [1/2]\log(1 + q_s\gamma_0) \approx [q_s\gamma_0]/[2\ln(2)]$ also decreases linearly with P_s. However, by comparing the function values, it can be easily shown that $C_{\mathrm{BF}} > C_{\mathrm{DT}}$. In addition, $C_{\mathrm{OAF}} \approx [1/2]\log(1 + q_s\gamma_0) \approx [q_s\gamma_0]/[2\ln(2)]$ for low power regions. Therefore, $C_{\mathrm{BF}} > C_{\mathrm{DT}}$ for low source and relay powers and the capacity in (3.21) is achieved by the NAF-BF scheme. Although the NAF-BF scheme is optimal, it does not provide great advantages in this scenario as $C_{\mathrm{BF}} \approx C_{\mathrm{OAF}} \approx C_{\mathrm{DT}}$.

3.4.2 Global Power Constraint

In the joint power constraint scenario, it can be seen from Theorem 3.2 that one must compare C_{DT} and C_{BF} in (3.28). We again include the OAF protocol in this comparison. Consider the following two cases.

Table 3.1 Capacity achieving protocols in high and low power regions

Constraint	Scenario	Capacity achieving protocol
Individual	High P_s, fixed P_r	DT
	High P_r, fixed P_s	NAF-BF ($q_s\gamma_0^2 < 4\gamma_1$) or DT ($q_s\gamma_0^2 > 4\gamma_1$)
	High P_s, P_r	DT
	Low P_s, P_r	NAF-BF $\approx$ DT $\approx$ OAF
Joint	High P_t	DT
	Low P_t	DT $\approx$ NAF-BF $\approx$ OAF

3.4.2.1 High Global Power

In this case, we assume that $P_t \to \infty$. Substituting (3.26) in (3.3), the capacity of the DT scheme is given as $C_{\text{DT}} = \log(1 + [q_t\gamma_0]/2) \approx \log(P_t)$, which grows logarithmically with P_t. For the NAF-BF scheme, $C_{\text{BF}} \approx [1/2]\log(P_t)$. This is again due to the fact that only one information symbol is sent in each cooperative frame. Likewise, $C_{\text{OAF}} \approx [1/2]\log(q_t)$. Since $C_{\text{BF}} > C_{\text{OAF}}$ from Theorem 3.2, $C_{\text{DT}} > C_{\text{BF}} > C_{\text{OAF}}$ for high values of q_t and the DT scheme achieves the capacity in (3.28).

3.4.2.2 Low Global Power

We then consider the case with $P_t \to 0$. In low transmission powers, $C_{\text{DT}} = \log(1+[q_t\gamma_0]/2) \approx [q_t\gamma_0]/[2\ln(2)]$. For the NAF-BF scheme, it can be shown from (3.25) that $q_{1,\text{BF}}, q_{2,\text{BF}} \to q_t/2$. Substituting (3.25) in (3.3) with $q_{1,\text{BF}} = q_{2,\text{BF}} = q_t/2$, the capacity of the NAF-BF scheme can be approximated as $C_{\text{BF}} \approx [1/2]\log(1 + q_t\gamma_0) \approx [q_t\gamma_0]/[2\ln(2)]$. For the OAF scheme, $q_{1,\text{OAF}} = q_t$. Substituting (3.27) in (3.3) with $q_{1,\text{OAF}} = q_t$, $C_{\text{OAF}} = [1/2]\log(1+q_t\gamma_0) \approx [q_t\gamma_0]/[2\ln(2)]$. Given that $\log(1+[q_t\gamma_0]/2) > [1/2]\log(1+q_t\gamma_0)$, $C_{\text{DT}} > C_{\text{BF}}$. Therefore, $C_{\text{DT}} > C_{\text{BF}} > C_{\text{OAF}}$ for low values of q_t and the capacity in (3.28) is again achieved by the DT scheme. The gain provided by DT is however not significant as $C_{\text{DT}} \approx C_{\text{BF}} \approx C_{\text{OAF}}$ in low power regions.

The advantages of the considered relaying strategies subject to different power constraints are summarized in Table 3.1.

3.5 Illustrative Examples

Based on the above analysis, this section illustrates the capacity and optimal power allocation for some specific network configurations. In particular, we shall consider two network models, namely the symmetric and the linear network model. Without loss of generality, we set $q_s = 2$, $z_r = 2$, and $q_t = 2$. Furthermore, we assume that the noise variances at both receiving nodes are unity, i.e., $N_d = N_r = N_0 = 1$.

3.5.1 Symmetric Network Model

Consider the symmetric network configuration where the source, relay and destination are all equidistant from each other. In this configuration, the nodes form an equilateral triangle and $\alpha_0 = \alpha_1 = \alpha_2 = \alpha$. Figure 3.3 shows the achievable rates of various AF protocols for the individual power constraint scenario with $\alpha = 1$. In particular, we consider the rates of the DT, NAF-FR, NAF-BF and OAF schemes using the power allocations in (3.13), (3.19), (3.16), and (3.14), respectively. For comparison purposes, we also consider the rate of the NAF system using a diagonal covariance matrix $q = 0$ and with full power at source and relay $\boldsymbol{q} = [q_s/2, q_s/2, z_r]$, denoted as the NAF-D protocol. In Fig. 3.3, the rates are normalized by the rate of the DT scheme (3.13), and are plotted against P_s/N_0 (in dB) for a fixed relay power $P_r/N_0 = 0$ dB. First, observe from Fig. 3.3 that only the NAF-BF, the NAF-FR, or the DT scheme can be optimal. This is in agreement with the capacity in (3.21), shown with square markers in Fig. 3.3 and all subsequent figures. As discussed in Sect. 3.4.1, it can be seen from Fig. 3.3 that AF relaying is not useful in high source power ranges as $C_{\mathrm{FR}}/C_{\mathrm{DT}} \to 1$ from below, $C_{\mathrm{BF}}/C_{\mathrm{DT}} \to 1/2$ and $C_{\mathrm{OAF}}/C_{\mathrm{DT}} \to 1/2$.

Figure 3.4 shows the rates of the considered AF protocols against P_r/N_0 for a fixed P_s/N_0 and still for the individual power constraint scenario. Two source powers are considered in Fig. 3.4: $P_s/N_0 = 0$ dB and $P_s/N_0 = 4.77$ dB. The rates in Fig. 3.4 are normalized by the rate of the DT scheme with $P_s/N_0 = 0$ dB, i.e., $C_{\mathrm{DT}} = 1$. Note from Fig. 3.4 that for $P_s/N_0 = 0$ dB, the capacity in (3.21) is achieved by either the NAF-FR scheme or the NAF-BF protocol. On the other hand, the capacity for $P_s/N_0 = 4.77$ dB is achieved by the DT scheme in the entire range of P_r/N_0 in Fig. 3.4. In both cases, the rate achieved by the OAF scheme approaches that of the NAF-BF protocol for high relay powers. This is in agreement

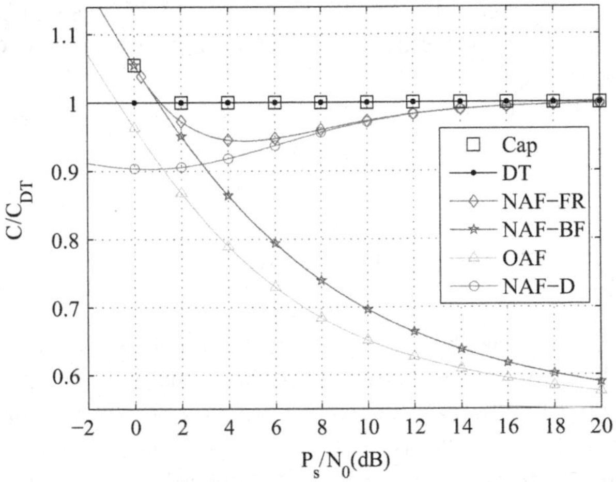

Fig. 3.3 Normalized rates of different protocols against P_s/N_0 ($P_r/N_0 = 0$ dB)

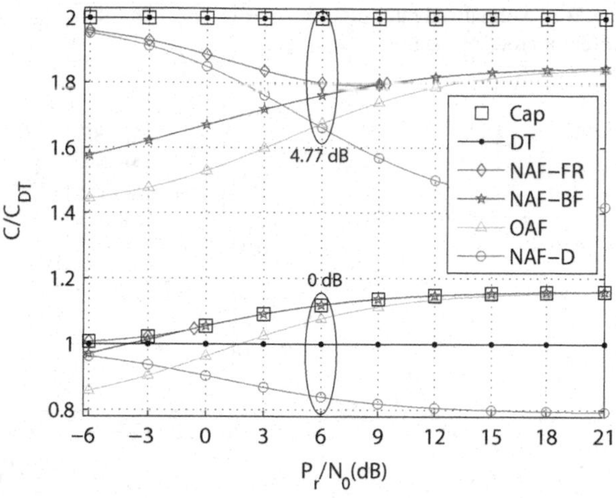

Fig. 3.4 Normalized rates of different protocols against P_r/N_0 ($P_s/N_0 = 0, 4.77$ dB)

with the large relay power analysis in Sect. 3.4.1 as $C_{\text{BF}} > C_{\text{DT}}$ for $P_s = 0/N_0$ dB (since $q_s\gamma_0^2 = 2 < 4\gamma_1 = 4$) and $C_{\text{DT}} > C_{\text{BF}}$ for $P_s/N_0 = 4.77$ dB (since $q_s\gamma_0^2 = 18 > 4\gamma_1 = 12$). It is also important to note from Figs. 3.3 and 3.4 that under certain configurations, using the relay might in fact reduce the achievable rate of the system.

The achievable rates over the symmetric network model are shown in Fig. 3.5 for the global constraint scenario. In this scenario, we consider the DT, NAF-BF, and OAF protocols with the power allocations in (3.26), (3.25), and (3.27), respectively. We also consider the OAF scheme with equal power allocation ($q = 0$ and $\boldsymbol{q} = [q_t/2, 0, q_t/2]$), denoted as OAF-EQ, and the NAF system using a diagonal covariance and equal power allocation ($q = 0$ and $\boldsymbol{q} = [q_t/3, q_t/3, q_t/3]$), denoted as NAF-EQ. As before, the rates in Fig. 3.5 are normalized by the rate of the DT scheme (3.26) and are plotted against P_t/N_0. Observe from Fig. 3.5 that for high transmission powers, $C_{\text{BF}}/C_{\text{DT}} \to 1/2$ and $C_{\text{OAF}}/C_{\text{DT}} \to 1/2$, which agrees with Sect. 3.4.2. Note also from Fig. 3.5 and Sect. 3.4.2 that $C_{\text{DT}} \approx C_{\text{BF}} \approx C_{\text{OAF}}$ in low power regions. More importantly, the DT scheme can be seen to be optimal for the entire range of P_t/N_0 in Fig. 3.5. This is because over the symmetric network configuration, it can be easily shown that $C_{\text{DT}} > C_{\text{BF}}$ regardless of the value of α.

3.5.2 Linear Network Model

We now consider the more practical linear network model which captures path-loss across transmission links. In this configuration, it is assumed that the relay is placed in the line between the source and the destination, the S-D distance is normalized to one, while the S-R and R-D distances are given by d and $(1-d)$, respectively. As a result, $\alpha_0 = 1$, $\alpha_1 = 1/d^\nu$ and $\alpha_2 = 1/(1-d)^\nu$, where ν is the path-loss

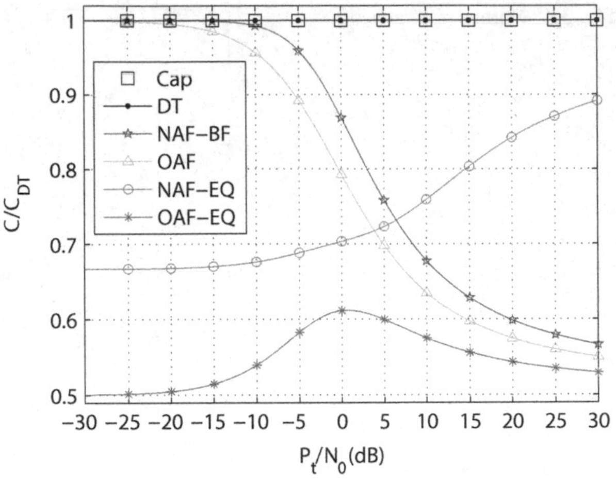

Fig. 3.5 Normalized rates of different protocols against P_t/N_0

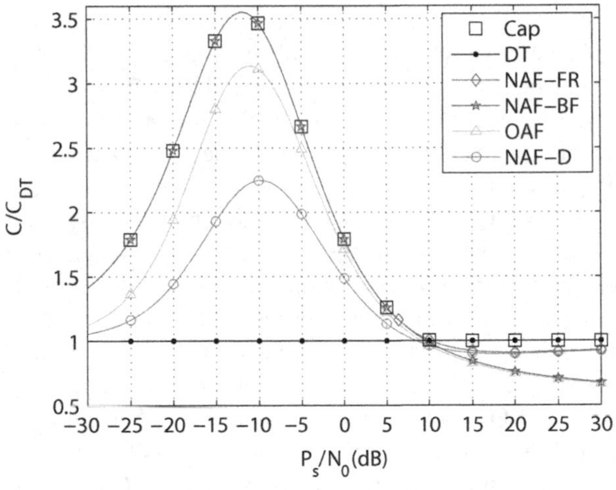

Fig. 3.6 Normalized rates of different protocols against $P_s/N_0 = P_r/N_0$ $(d = 0.5)$

exponent, which is set to 3 in this section for convenience. Figure 3.6 shows the normalized rates of the per-node constrained systems for $d = 0.5$ and $P_s/N_0 = P_r/N_0$. First, note from Fig. 3.6 that the capacity in Theorem 3.1 is achieved by either the NAF-BF, the NAF-FR, or the DT scheme. As explained in Sect. 3.4.1, it can be seen from Fig. 3.6 that $C_{\mathrm{FR}}/C_{\mathrm{DT}} \to 1$ and $C_{\mathrm{BF}}/C_{\mathrm{DT}}, C_{\mathrm{OAF}}/C_{\mathrm{DT}} \to 1/2$ when $P_s, P_r \to \infty$, whereas $C_{\mathrm{BF}}/C_{\mathrm{DT}}, C_{\mathrm{OAF}}/C_{\mathrm{DT}} \to 1$ when $P_s, P_r \to 0$, i.e., AF relaying does not provide significant gains in low and high power regions. However, in medium power ranges, the NAF-BF system can be seen to provide impressive gains over the DT scheme. Specifically, the rate of the NAF-BF protocol in Fig. 3.6 is 3.5 times that of the DT scheme at $P_s/N_0 = -10\,\mathrm{dB}$.

Figure 3.7 shows the rates of the individual power constrained systems against P_r/N_0 for fixed $P_s/N_0 = 0\,\mathrm{dB}$. Three values of d are considered: $d = 0.1, 0.5$ and 0.9. Note that the rate of the DT scheme is equal for these values as $\alpha_0 = 1$. Observe

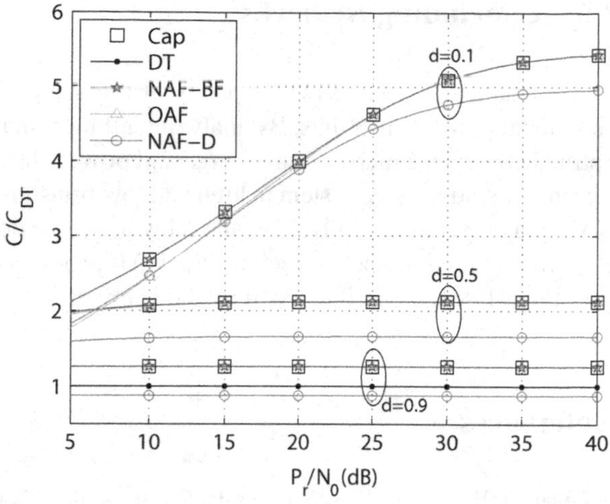

Fig. 3.7 Normalized rates of different protocols against P_r/N_0 ($P_s/N_0 = 0\,\text{dB}$, $d = 0.1, 0.5, 0.9$)

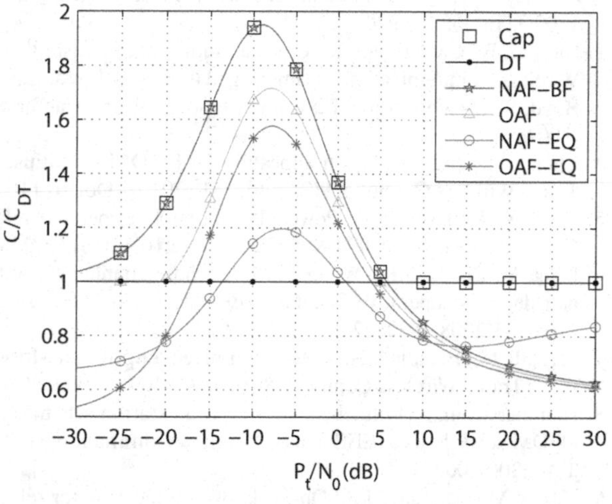

Fig. 3.8 Normalized rates of different protocols against P_t/N_0 ($d = 0.6$)

from Fig. 3.7 that the rate of the OAF system approaches that of the NAF-BF scheme for high relay powers, as expected. Furthermore, $C_{\text{BF}} > C_{\text{DT}}$ for all values of d in Fig. 3.7 since $q_s\gamma_0^2 < 4\gamma_1$. More importantly, the asymptotic gain offered by the NAF-BF scheme over the DT system increases as the relay gets closer to the source. Specifically, the NAF-BF system presents asymptotic gains of 1.2, 2.1 and 5.4 times the rate of the DT protocol for $d = 0.9$, 0.5 and 0.1, respectively.

The rates of the global power constrained systems are shown in Fig. 3.8 for $d = 0.6$. Note that the asymptotic behavior described in Sect. 3.4.2 holds for all systems in Fig. 3.8. Although AF relaying does not provide any advantage over DT for high and low transmission powers, it can be seen from Fig. 3.8 that significant gains can be achieved in medium power levels. For instance, an almost two-fold increase in rate over the DT protocol is attained by the NAF-BF system at $P_t = -10\,\text{dB}$.

3.6 Concluding Remarks

In this chapter, we derived the capacity of the static NAF system under individual and global power constraints. By analyzing all local maximizers, expressions for the optimal covariance matrix at the source and power allocation at relay were obtained. By further studying the system in high and low transmission powers, we showed that NAF relaying is useful only when the relay has large power and the source does not. However, in medium power regions, the NAF protocol can provide significant gains over the DT scheme as illustrated by the symmetric and linear network models.

References

1. Azarian, K., Gamal, H.E., Schniter, P.: On the achievable diversity-multiplexing tradeoff in half-duplex cooperative channels. IEEE Trans. Inf. Theory **51**(12), 4152–4172 (2005). Doi:10.1109/TIT.2005.858920
2. Biglieri, E., Calderbank, R., Constantinides, A., Goldsmith, A., Paulraj, A., Poor, H.: MIMO Wireless Communications. Cambridge University Press, Cambridge (2007)
3. Boyd, S., Vandenberghe, L.: Convex Optimization. Cambridge University Press, Cambridge (2004)
4. Chen, M., Yener, A.: Power allocation for F/TDMA multiuser two-way relay networks. IEEE Trans. Wireless Commun. **9**(2), 546–551 (2010). Doi:10.1109/TWC.2010.02.090336
5. Deng, X., Haimovich, A.: Power allocation for cooperative relaying in wireless networks. IEEE Commun. Lett. **9**(11), 994–996 (2005). Doi:10.1109/LCOMM.2005.11012
6. Ding, Y., Zhang, J.K., Wong, K.M.: Ergodic channel capacities for the amplify-and-forward half-duplex cooperative systems. IEEE Trans. Inf. Theory **55**(2), 713–730 (2009). Doi:10.1109/TIT.2008.2009822
7. Farhadi, G., Beaulieu, N.: Power-optimized amplify-and-forward multi-hop relaying systems. IEEE Trans. Wireless Commun. **8**(9), 4634–4643 (2009). Doi:10.1109/TWC.2009.080987
8. Hammerstrom, I., Wittneben, A.: Power allocation schemes for amplify-and-forward MIMO-OFDM relay links. IEEE Trans. Wireless Commun. **6**(8), 2798–2802 (2007). Doi:10.1109/TWC.2007.06071
9. Hasna, M., Alouini, M.S.: Optimal power allocation for relayed transmissions over Rayleigh-fading channels. IEEE Trans. Wireless Commun. **3**(6), 1999–2004 (2004). Doi:10.1109/TWC.2004.833447
10. Jang, Y.U., Jeong, E.R., Lee, Y.: A two-step approach to power allocation for OFDM signals over two-way amplify-and-forward relay. IEEE Trans. Signal Process. **58**(4), 2426–2430 (2010). Doi:10.1109/TSP.2010.2040415
11. Jiménez-Rodríguez, L., Tran, N.H., Le-Ngoc, T.: On the capacity of the static half-duplex non-orthogonal AF relay channel. IEEE Trans. Wireless Commun. **13**(2), 1034–1046 (2014). Doi:10.1109/TWC.2013.010214.130868
12. Jingmei, Z., Qi, Z., Chunju, S., Ying, W., Ping, Z., Zhang, Z.: Adaptive optimal transmit power allocation for two-hop non-regenerative wireless relaying system. In: Proceedings of IEEE Vehicular Technology Conference, vol. 2, pp. 1213–1217 (2004). Doi:10.1109/VETECS.2004.1389025
13. Ko, Y., Ardakani, M., Vorobyov, S.: Power allocation strategies across N orthogonal channels at both source and relay. IEEE Trans. Commun. **60**(6), 1469–1473 (2012). Doi:10.1109/TCOMM.2012.040212.100193

14. Li, Y., Vucetic, B., Zhou, Z., Dohler, M.: Distributed adaptive power allocation for wireless relay networks. IEEE Trans. Wireless Commun. **6**(3), 948–958 (2007). Doi:10.1109/TWC.2007.05256
15. Nabar, R.U., Bölcskei, H., Kneubühler, F.W.: Fading relay channel: Performance limits and space-time signal design. IEEE J. Sel. Areas Commun. **22**(6), 1099–1109 (2004). Doi:10.1109/JSAC.2004.830922
16. Pischella, M., Le Ruyet, D.: Optimal power allocation for the two-way relay channel with data rate fairness. IEEE Commun. Lett. **15**(9), 959–961 (2011). Doi:10.1109/LCOMM.2011.070711.110789
17. Saito, M., Athaudage, C.R.N., Evans, J.: On power allocation for dual-hop amplify-and-forward OFDM relay systems. In: Proceedings of IEEE Global Telecommunications Conference, pp. 1–6 (2008). Doi:10.1109/GLOCOM.2008.ECP.789
18. Sidhu, G., Gao, F., Chen, W., Nallanathan, A.: A joint resource allocation scheme for multiuser two-way relay networks. IEEE Trans. Commun. **59**(11), 2970–2975 (2011). Doi:10.1109/TCOMM.2011.071111.100199
19. Telatar, I.E.: Capacity of multi-antenna Gaussian channels. European Trans. Telecommun. **10**(6), 585–595 (1999)
20. Zhao, Y., Adve, R., Lim, T.: Improving amplify-and-forward relay networks: Optimal power allocation versus selection. IEEE Trans. Wireless Commun. **6**(8), 3114–3123 (2007). Doi:10.1109/TWC.2007.06026

Chapter 4
Half-Duplex AF Relaying: Achievable Rate and Power Allocation Over Fading Channels

In the previous chapter, we studied the capacity of the static NAF system in which the transmitted codeword spans over a single channel realization. Such analysis involved the maximization of the conditional mutual information. As discussed in the introduction of Chap. 3, the capacities for the DHAF, OAF and TWAF protocols over a static environment have all been addressed in the literature. On the other hand, when the transmitted codeword spans over several realizations, the ergodic achievable rate becomes the criterion of interest. In this fading environment, the capacity study involves two steps. At first, one needs to compute the ergodic achievable rate by averaging the conditional mutual information in (2.5), (2.9), (2.11), and (2.14) over the channel realizations $\boldsymbol{h}$. This expectation will be different depending on the selected amplification coefficient, e.g., for either the FG or the CI coefficient. Optimization must then be applied on the derived expressions to find the optimal average power allocation schemes that maximize the achievable rate or sum rate for OW or TW AF systems, respectively.

Several works in the literature have investigated achievable rate expressions for HD AF systems over fast fading channels. For instance, expressions under different assumptions have been derived in [2, 3, 8, 16, 18] for the DHAF-CI system, in [13, 17, 19, 22] for the DHAF-FG, in [7] for the OAF-CI, in [4, 7, 9] for the OAF-FG, and in [6, 10, 20, 21] for TWAF-CI. In addition, power allocation schemes for TWAF-CI were proposed in [10, 21] assuming equal power at the source nodes. As discussed in Chap. 3, the capacity of the FG- and CI-NAF channel was investigated in [5], where it was shown that a Gaussian input with a diagonal covariance matrix maximizes the achievable rate. However, the computation of the rate and power allocation that achieves the capacity was performed numerically in [5]. Interestingly, by examining the rate in high SNR regimes, it was demonstrated in [5] that NAF relaying is not useful compared to the DT scheme. This result, however, is only applicable to the FG system over unit variance channels.

L. Jiménez Rodríguez et al., *Amplify-and-Forward Relaying in Wireless Communications*, SpringerBriefs in Computer Science,
DOI 10.1007/978-3-319-17981-0_4

In most of the above works, the achievable rates are obtained in terms of complicated mathematical functions. Although many of these functions provide an accurate representation so that Monte Carlo simulations are no longer needed, they do not provide insightful expressions of the achievable rate. In particular, most of the above works resort to plotting when discussing the behavior of the considered AF system. As a result, it is very difficult to obtain optimal average power allocation strategies based on channel statistics to further improve the achievable rates or sum rates of OW and TW AF systems. To the best of our knowledge, analyses on optimal power allocation schemes over fading channels are scarce in the literature and numerical optimization is usually needed. In addition, the techniques proposed in the current literature are applicable to either OW or TW relaying using either the CI or the FG amplification coefficient. As such, it is not clear what the effect of using the FG and CI coefficients is to the performance of different AF protocols.

Inspired by the above observations, this chapter proposes a general approach to study ergodic achievable rates and power allocation schemes for HD single-relay AF systems over Rayleigh fading channels. Our main idea is to rely on the capacity of a two-branch maximal-ratio combining (MRC) system [14, 15] and on a simple approximation to the logarithm to derive tight approximations to the achievable rate in high and low transmission power regions. Different from previous works, the proposed approach is applicable to OW and TW protocols using either the CI or the FG coefficient. Furthermore, the proposed approach results in insightful expressions from which comparisons among different coefficients, and optimal average power allocations can be obtained. Specifically, approximations are first derived for the DHAF, NAF and two-phase TW relaying schemes using both the CI and FG coefficients. The derived approximations are tight in their respective power ranges and easy to analyze, as they are given in terms of the well-known exponential integral. The systems using the FG and CI coefficients are then compared. While the CI technique is better for DH and TW schemes at high powers, the FG scheme is superior in low powers for OW systems and in high powers for the NAF protocol. Asymptotic power allocation schemes based on the derived approximations are then proposed to maximize the rate or sum rate for OW and TW systems, respectively. Finally, for the specific case of the DH scheme with CI, a closed-form expression of the rate is derived using the MRC approach and bisection on the closed-form derivative is proposed to obtain the optimal power allocation.

4.1 Problem Formulation

In this chapter, we consider the fast fading scenario in which the coherence time of the channel is small enough so that the transmitted codeword spans several realizations. Specifically, Rayleigh block fading is assumed and thus the channel gains among the nodes are distributed as $h_l \sim \mathscr{CN}(0, \phi_l)$, $l \in \{0, 1, 2\}$. The variances $\boldsymbol{\Phi} = [\phi_0, \phi_1, \phi_2]$ account for different pathloss effects over the links. Recall that $\alpha_l = |h_l|^2$. Since all gains are complex circular Gaussian, α_l is exponentially distributed with mean ϕ_l. In this ergodic environment, the channel gains are rapidly changing and thus only receiver CSI is considered. In particular, the destination node has full CSI, whereas the relay has either statistical or full knowledge of the incoming links. Two types of amplification coefficients can then be used at the relay, namely, the FG and the CI coefficients with $\Xi = \phi$ and $\Xi = \alpha$ in (2.4) and (2.13). Both the OW and the TW relay protocols described in Sect. 2.2 are considered. For OW relaying, the input-output relation of the NAF channel is given by (2.1). This protocol includes the OAF with $q_2 = 0$ in (2.8) and the DT scheme in (2.6) with $z_2 = 0$ as special cases. Moreover, the DHAF scheme in (2.10) is obtained when the direct link is blocked $\phi_0 \approx 0$. For TW relaying, we consider the two-phase TW protocol with input-output relation given by (2.12).

For a given amplification coefficient and power allocation scheme $\boldsymbol{q} = [q_1, q_2, z_2]$, it was shown in [5] that the unconditional mutual information between the input and output of the NAF channel $I = \mathbb{E}[I|\boldsymbol{h}]$ is maximized by using Gaussian inputs with a diagonal covariance matrix $\boldsymbol{Q}$. The achievable rate conditioned on $\boldsymbol{h}$ is then given as (2.5) with $q_{12} = 0$. For TW relaying, the conditional rate in the $S_k \to S_i$ direction ($i \in \{1, 2\}, k = 3 - i$) is given by (2.14). The unconditional achievable rate is then written as $I_i = \mathbb{E}[I_i|\boldsymbol{h}]$, and the unconditional achievable sum rate is $I_{\text{sum}} = I_1 + I_2$. The objective of this chapter is to maximize the unconditional achievable rate or sum rate for different OW or TW relay protocols, respectively. Similar to previous works [10, 21], we focus on the global constraint scenario. Let $P_s = P_r = P_t$ in (2.1), $P_{s1} = P_{s2} = P_t$ in (2.12) and $q_1 + q_2 + z_2 \leq q_t$. The system is then allowed to spend an average power of up to $q_t P_t$ per transmission frame. To have an average power constraint of P_t per symbol period, the value of q_t can be set to the number of phases required to communicate from S_1 to S_2 in OW relaying, or to exchange symbols between the two source nodes in TW schemes. In the case of OW relaying, we consider the maximum ergodic achievable rate, i.e., the power-constrained capacity, as the performance criterion. Specifically, the capacity can be calculated as

$$C = \max_{q_1, q_2, z_2 \geq 0} \quad \mathbb{E}[I|\boldsymbol{h}] \quad \text{s.t.} \quad q_1 + q_2 + z_2 \leq q_t, \tag{4.1}$$

where the conditional mutual information $I|\boldsymbol{h}$ is given by (2.5), (2.9), or (2.11) depending on the protocol under consideration. For TW systems, our objective is to maximize the achievable sum rate, which can be expressed as

$$C_{\text{sum}} = \max_{q_1, q_2, z_2 \geq 0} \quad \mathbb{E}[I_1|\boldsymbol{h} + I_2|\boldsymbol{h}] \quad \text{s.t.} \quad q_1 + q_2 + z_2 \leq q_t, \tag{4.2}$$

where $I_i|\boldsymbol{h}$ is given in (2.14). For both problems in (4.1) and (4.2), the CI and FG techniques will be addressed.

Finding the solution to the optimization problems in (4.1) and (4.2) involves two steps. At first, one needs to compute the unconditional achievable rates for a given power allocation scheme. Such rates then need to be optimized with respect to $\boldsymbol{q}$. In the following, we present solutions to these problems by focusing on the high and low transmission power regimes, i.e., high and low values of P_t.

4.2 Achievable Rates and Closed-Form Approximations

In this section, we derive approximations to the achievable rates and provide comparisons between the CI and FG techniques. As shall be shown in Sects. 4.3 and 4.5, the approximations are tight in their respective power ranges and provide important insights on the problems in (4.1) and (4.2). To this end, we first outline a general approach that can be applied to analyze the rates of the considered AF protocols. The following proposition, which is based on the capacity of a MRC system [14, 15] with two branches, will be used to examine the rates.

Proposition 4.1. *Let $a_0 > 0$ be a constant, and ω_1 and ω_2 be independent exponentially distributed random variables with means μ_1 and μ_2, respectively. Define*

$$\mathscr{J}(x) = \exp(x)E_1(x), \tag{4.3}$$

where $E_1(\cdot)$ is the exponential integral [1]:

$$E_1(x) = \int_x^{\infty} \frac{e^{-u}}{u} du = -\left(\gamma + \ln(x) + \sum_{n=1}^{\infty} \frac{(-1)^n x^n}{n!n}\right), \tag{4.4}$$

and γ is the Euler constant. Then, for $a_0 > 0$, one has

$$\mathbb{E}[\ln(a_0 + \omega_1)] = \ln(a_0) + \mathscr{J}(a_0/\mu_1), \tag{4.5a}$$

$$\mathbb{E}[\ln(a_0 + \omega_1 + \omega_2)] = \begin{cases} 1 + \ln(a_0) + \left(1 - \frac{a_0}{\mu_1}\right) \mathscr{J}\left(\frac{a_0}{\mu_1}\right), & \mu_1 = \mu_2 \\ \ln(a_0) + \frac{\mu_1 \mathscr{J}(a_0/\mu_1) - \mu_2 \mathscr{J}(a_0/\mu_2)}{\mu_1 - \mu_2}, & \mu_1 \neq \mu_2, \end{cases} \tag{4.5b}$$

and for $a_0 = 0$,

$$\mathbb{E}[\ln(\omega_1)] = \ln(\mu_1) - \gamma, \tag{4.6a}$$

$$\mathbb{E}[\ln(\omega_1 + \omega_2)] = \begin{cases} 1 - \gamma + \ln(\mu_1), & \mu_1 = \mu_2 \\ \frac{\mu_1 \ln(\mu_1) - \mu_2 \ln(\mu_2)}{\mu_1 - \mu_2} - \gamma, & \mu_1 \neq \mu_2. \end{cases} \tag{4.6b}$$

Proof. Factoring a_0, (4.5a) and (4.5b) are simply the capacities of a single-input single-output system [15, Eq. (15.26)], and of a two-branch MRC combiner with equal [15, Eq. (15.33)] and unequal [14] average SNRs. Equations (4.6a) and (4.6b) then follow from the fact that $\lim_{a_0 \to 0} \mathscr{J}(a_0/\mu) = -\gamma - \ln(a_0) + \ln(\mu)$ [12]. □

From the above proposition, approximations to the achievable rate in high power regions can be derived as follows. First, the conditional rate of the considered AF systems can be written as $\log(1+f(\boldsymbol{h}))$, where $f(\boldsymbol{h})$ can be represented as a fraction of two functions, i.e., $f(\boldsymbol{h}) = f_1(\boldsymbol{h})/f_2(\boldsymbol{h})$. The achievable rate can then be written as $\mathbb{E}[\log(1+f(\boldsymbol{h}))] = \mathbb{E}[\log(f_1(\boldsymbol{h})+f_2(\boldsymbol{h}))] - \mathbb{E}[\log(f_2(\boldsymbol{h}))]$. At high transmission powers, one can ignore the lower order terms of $f_1(\boldsymbol{h}) + f_2(\boldsymbol{h})$ and $f_2(\boldsymbol{h})$. The expectations will have a form similar to either (4.5) or (4.6). The approximation to the achievable rate at high powers is then obtained by ignoring the lower order terms and using Proposition 4.1 to take the expectation of the remaining terms. At low transmission powers, the conditional rate can be approximated as $\mathbb{E}[\log(1 + f(\boldsymbol{h}))] \approx \mathbb{E}[f(\boldsymbol{h})]/\ln(2)$, where the fact that $\ln(1+x) \approx x$ for small $x > 0$ has been used. One can then easily take the expectation of $f(\boldsymbol{h})$ by ignoring the higher order terms.

Based on the above approaches, in the following, we shall analyze the achievable rates for each of the considered AF systems in further detail. Without loss of generality, we assume unit noise power at all nodes $N_d = N_r = N_{di} = N_0 = 1$.

4.2.1 OW DHAF Systems

To illustrate the proposed approach, consider first the DH system using the coefficients in (2.4). As previously mentioned, the direct $S_1 - S_2$ is under heavy shadowing in the DH protocol. Substituting (2.4) into (2.11), the conditional achievable rate for this system simplifies to

$$I_{\text{DH}}|\boldsymbol{h} = \frac{1}{2}\log\left(1 + \frac{P_t^2 q_1 z_2 \alpha_1 \alpha_2}{P_t(q_1 \Xi_1 + z_2\alpha_2) + 1}\right) = \frac{1}{2}\log\left[1 + f_{\text{DH}}(\boldsymbol{h})\right], \tag{4.7}$$

where $\Xi_1 = \phi_1$ for the FG coefficient and $\Xi_1 = \alpha_1$ for the CI one. As outlined before, in this case $f_{\text{DH}}(\boldsymbol{h}) = f_{1,\text{DH}}(\boldsymbol{h})/f_{2,\text{DH}}(\boldsymbol{h})$ with $f_{1,\text{DH}}(\boldsymbol{h}) = P_t^2 q_1 z_2 \alpha_1 \alpha_2$ and $f_{2,\text{DH}}(\boldsymbol{h}) = P_t(q_1\Xi_1 + z_2\alpha_2) + 1$. At high powers,

$$\begin{aligned} f_{1,\text{DH}}(\boldsymbol{h}) + f_{2,\text{DH}}(\boldsymbol{h}) &= P_t^2 q_1 z_2 \alpha_1 \alpha_2 + O(P_t), \\ f_{2,\text{DH}}(\boldsymbol{h}) &= P_t(q_1\Xi_1 + z_2\alpha_2) + O(1). \end{aligned}$$

The achievable rate in (4.7) can then be written in high power regions as

$$I_{\text{DH}}|\boldsymbol{h} = \frac{1}{2}\log\left(\frac{f_{1,\text{DH}}(\boldsymbol{h}) + f_{2,\text{DH}}(\boldsymbol{h})}{f_{2,\text{DH}}(\boldsymbol{h})}\right) = \frac{1}{2}\log\left(\frac{P_t^2 q_1 z_2 \alpha_1 \alpha_2 + O(P_t)}{P_t(q_1\Xi_1 + z_2\alpha_2) + O(1)}\right). \tag{4.8}$$

The rate for the FG system can be approximated from (4.8) by ignoring the lower order terms and substituting $\Xi_1 = \phi_1$ as

$$\begin{aligned} I_{\text{DH}}^{\text{FG}} &\approx \frac{1}{2}\log(P_t q_1 z_2) + \frac{1}{2\ln(2)}\{\mathbb{E}[\ln(\alpha_1)] + \mathbb{E}[\ln(\alpha_2)] - \mathbb{E}[\ln(q_1\phi_1 + z_2\alpha_2)]\} \\ &= \frac{1}{2}\log(P_t) + \log(\sqrt{z_2\phi_2}) - \frac{1}{2\ln(2)}\mathcal{J}\left(\frac{q_1\phi_1}{z_2\phi_2}\right) - \frac{\gamma}{\ln(2)}, \end{aligned} \tag{4.9}$$

where the first two expectations are solved using (4.6a), and the last one using (4.5a) in Proposition 4.1. Similarly, by substituting $\Xi_1 = \alpha_1$, the rate of the CI system can be approximated at high powers from (4.8) and Proposition 4.1 as

$$\begin{aligned} I_{\mathrm{DH}}^{\mathrm{CI}} &\approx \frac{1}{2}\log(P_t q_1 z_2) + \frac{1}{2\ln(2)}\{\mathbb{E}[\ln(\alpha_1)] + \mathbb{E}[\ln(\alpha_2)] - \mathbb{E}[\ln(q_1\alpha_1 + z_2\alpha_2)]\} \\ &= \frac{1}{2}\log(P_t) + \begin{cases} \log(\sqrt{q_1\phi_1}) - \frac{\gamma}{2\ln(2)} - \frac{1}{2\ln(2)}, & q_1\phi_1 = z_2\phi_2 \\ \frac{q_1\phi_1\log(z_2\phi_2) - z_2\phi_2\log(q_1\phi_1)}{2(q_1\phi_1 - z_2\phi_2)} - \frac{\gamma}{2\ln(2)}, & q_1\phi_1 \neq z_2\phi_2, \end{cases} \end{aligned} \tag{4.10}$$

where (4.6a) and (4.6b) have been used. From (4.8), the rate difference between the FG and CI techniques can be approximated as

$$I_{\mathrm{DH}}^{\mathrm{FG}} - I_{\mathrm{DH}}^{\mathrm{CI}} \approx \frac{1}{2}\mathbb{E}\left[\log\left(\frac{q_1\alpha_1 + z_2\alpha_2}{q_1\phi_1 + z_2\alpha_2}\right)\right]. \tag{4.11}$$

Applying Jensen's inequality to α_1 in (4.11), it can be easily shown that $I_{\mathrm{DH}}^{\mathrm{FG}} - I_{\mathrm{DH}}^{\mathrm{CI}} \leq 0$. Therefore, CI is more beneficial than FG in high power regions.

In low power regions, as outlined before, the conditional achievable rate can be approximated from (4.7) as

$$I_{\mathrm{DH}}|\boldsymbol{h} \approx \frac{f_{\mathrm{DH}}(\boldsymbol{h})}{2\ln(2)} = \frac{P_t^2 q_1 z_2 \alpha_1 \alpha_2}{2\ln(2)[P_t(q_1\Xi_1 + z_2\alpha_2) + 1]}.$$

For both FG and CI systems, the unconditional achievable rate can then be approximated as

$$I_{\mathrm{DH}} \approx \mathbb{E}\left[\frac{P_t^2 q_1 z_2 \alpha_1 \alpha_2}{2\ln(2)[1 + O(P_t)]}\right] = \frac{P_t^2 q_1 z_2 \phi_1 \phi_2}{2\ln(2)}. \tag{4.12}$$

Hence, the rates of the CI and FG systems converge in low power regions and the performance is asymptotically independent of the selected coefficient. However, for slightly higher powers, it is straightforward to show that

$$I_{\mathrm{DH}}^{\mathrm{FG}} - I_{\mathrm{DH}}^{\mathrm{CI}} \approx \frac{1}{2\ln(2)} \cdot \mathbb{E}\left[f_{\mathrm{DH}}^{\mathrm{FG}}(\boldsymbol{h}) - f_{\mathrm{DH}}^{\mathrm{CI}}(\boldsymbol{h})\right] \geq 0, \tag{4.13}$$

as $f_{\mathrm{DH}}^{\mathrm{FG}}(\boldsymbol{h}) - f_{\mathrm{DH}}^{\mathrm{CI}}(\boldsymbol{h})$ is convex with respect to α_1. Therefore, FG is better than CI, and the CI system approaches the FG one from below at sufficiently low powers.

4.2.2 OW Cooperative Systems

Consider now the general cooperative NAF system in (2.5) with $q_{12} = 0$. As noted before, the OAF and DT schemes are special cases of this protocol. First, consider the NAF system with $q_1, q_2, z_2 > 0$. The conditional achievable rate of this system using the coefficients in (2.4) can be written as

$$I_{\text{NAF}}|\boldsymbol{h} = \frac{1}{2}\log\left[1 + f_{\text{NAF}}(\boldsymbol{h})\right], \tag{4.14}$$

where $f_{\text{NAF}}(\boldsymbol{h}) = f_{1,\text{NAF}}(\boldsymbol{h})/f_{2,\text{NAF}}(\boldsymbol{h})$ with

$$\begin{aligned} f_{1,\text{NAF}}(\boldsymbol{h}) =& P_t^3(q_1^2 q_2 \alpha_0^2 \Xi_1) + P_t^2 q_1(q_2\alpha_0^2 + q_1\alpha_0\Xi_1 + z_2\alpha_0\alpha_2 + z_2\alpha_1\alpha_2 + q_2\alpha_0\Xi_1) \\ &+ P_t\alpha_0(q_1 + q_2), \\ f_{2,\text{NAF}}(\boldsymbol{h}) =& P_t(q_1\Xi_1 + z_2\alpha_2) + 1. \end{aligned}$$

As in the DH case, high power approximations to the achievable rate of the FG and CI systems can be obtained from (4.14) and Proposition 4.1 by ignoring the lower order terms of $f_{1,\text{NAF}}(\boldsymbol{h})$ and $f_{2,\text{NAF}}(\boldsymbol{h})$ as

$$I_{\text{NAF}}|\boldsymbol{h} = \frac{1}{2}\log\left(\frac{P_t^3(q_1^2 q_2\alpha_0^2\Xi_1) + O(P_t^2)}{P_t(q_1\Xi_1 + z_2\alpha_2) + O(1)}\right) \approx \frac{1}{2}\log\left(\frac{P_t^3[q_1^2 q_2\alpha_0^2\Xi_1]}{P_t(q_1\Xi_1 + z_2\alpha_2)}\right).$$

Thus, by following the same procedure, the achievable rate for the FG system is approximated by setting $\Xi_1 = \phi_1$ and solving the expectations as

$$\begin{aligned} I_{\text{NAF}}^{\text{FG}} \approx& \frac{1}{2}\log(P_t^2 q_1^2 q_2\phi_1) + \frac{1}{2\ln(2)}\{2\cdot\mathbb{E}[\ln(\alpha_0)] - \mathbb{E}[\ln(q_1\phi_1 + z_2\alpha_2)]\} \\ =& \log(P_t) + \log(\phi_0\sqrt{q_1 q_2}) - \frac{1}{2\ln(2)}\mathscr{I}\left(\frac{q_1\phi_1}{z_2\phi_2}\right) - \frac{\gamma}{\ln(2)}, \end{aligned} \tag{4.15}$$

where (4.6a) and (4.5a) have been used. The rate of the CI system with $\Xi_1 = \alpha_1$ can be similarly approximated in high power regions as

$$I_{\text{NAF}}^{\text{CI}} \approx \frac{1}{2}\log(P_t^2 q_1^2 q_2) + \frac{1}{2\ln(2)}\{2\cdot\mathbb{E}[\ln(\alpha_0)] + \mathbb{E}[\ln(\alpha_1)] - \mathbb{E}[\ln(q_1\alpha_1 + z_2\alpha_2)]\}$$

$$= \log(P_t) + \begin{cases} \log(\phi_0\sqrt{q_1 q_2}) - \frac{\gamma}{\ln(2)} - \frac{1}{2\ln(2)}, & q_1\phi_1 = z_2\phi_2 \\ \log(\phi_0 q_1\sqrt{q_2\phi_1}) - \frac{q_1\phi_1\log(q_1\phi_1) - z_2\phi_2\log(z_2\phi_2)}{2(q_1\phi_1 - z_2\phi_2)} - \frac{\gamma}{\ln(2)}, & q_1\phi_1 \neq z_2\phi_2, \end{cases} \tag{4.16}$$

by applying (4.6a) and (4.6b). The rate difference between the FG and CI system for the NAF protocol can be written from (4.14) and Proposition 4.1 as

$$I_{\text{NAF}}^{\text{FG}} - I_{\text{NAF}}^{\text{CI}} \approx \frac{1}{2\ln(2)} \times \begin{cases} 1 - \mathscr{J}(1), & q_1\phi_1 = z_2\phi_2 \\ \frac{z_2\phi_2[\ln(q_1\phi_1)-\ln(z_2\phi_2)]}{q_1\phi_1 - z_2\phi_2} - \mathscr{J}\left(\frac{q_1\phi_1}{z_2\phi_2}\right), & q_1\phi_1 \neq z_2\phi_2. \end{cases} \tag{4.17}$$

As shown in [12], (4.17) is strictly positive and hence $I_{\text{NAF}}^{\text{FG}} \geq I_{\text{NAF}}^{\text{CI}}$ in high power regions. In low power regions, the rate for the FG and CI systems can be approximated following the proposed approach as

$$I_{\text{NAF}} \approx \mathbb{E}\left[\frac{f_{\text{NAF}}(\boldsymbol{h})}{2\ln(2)}\right] = \mathbb{E}\left[\frac{P_t\alpha_0[q_1+q_2] + O(P_t^2)}{2\ln(2)[1+O(P_t)]}\right] \approx \frac{P_t(q_1+q_2)\phi_0}{2\ln(2)}. \tag{4.18}$$

Furthermore, $f_{\text{NAF}}^{\text{FG}}(\boldsymbol{h}) - f_{\text{NAF}}^{\text{CI}}(\boldsymbol{h})$ can be shown to be a convex function with respect to α_1 and thus $I_{\text{NAF}}^{\text{FG}} \geq I_{\text{NAF}}^{\text{CI}}$. The FG system then outperforms the CI one in low powers, and the difference decreases as the power decreases.

Consider now the OAF protocol in which the source remains silent in the second phase, i.e., $q_2 = 0$. In this case, the conditional achievable rate in (2.9) simplifies to

$$I_{\text{OAF}}|\boldsymbol{h} = \frac{1}{2}\log\left(1 + P_t q_1\alpha_0 + \frac{P_t^2 q_1 z_2\alpha_1\alpha_2}{P_t(q_1\Xi_1 + z_2\alpha_2) + 1}\right) = \frac{1}{2}\log\left[1 + f_{\text{OAF}}(\boldsymbol{h})\right], \tag{4.19}$$

where $f_{\text{OAF}}(\boldsymbol{h}) = f_{1,\text{OAF}}(\boldsymbol{h})/f_{2,\text{OAF}}(\boldsymbol{h})$ with

$$\begin{aligned} f_{1,\text{OAF}}(\boldsymbol{h}) &= P_t^2 q_1(q_1\alpha_0\Xi_1 + z_2\alpha_0\alpha_2 + z_2\alpha_1\alpha_2) + P_t q_1\alpha_0 \\ &= P_t^2 q_1(q_1\alpha_0\Xi_1 + z_2\alpha_0\alpha_2 + z_2\alpha_1\alpha_2) + O(P_t), \\ f_{2,\text{OAF}}(\boldsymbol{h}) &= P_t(q_1\Xi_1 + z_2\alpha_2) + 1 \\ &= P_t(q_1\Xi_1 + z_2\alpha_2) + O(1). \end{aligned}$$

At high transmission powers, the unconditional rate can be approximated by ignoring the lower order terms as

$$I_{\text{OAF}} \approx \frac{1}{2}\log(P_t) + \frac{1}{2}\mathbb{E}\left[\log\left(\frac{q_1(q_1\alpha_0\Xi_1 + z_2\alpha_0\alpha_2 + z_2\alpha_1\alpha_2)}{q_1\Xi_1 + z_2\alpha_2}\right)\right].$$

Unfortunately, due to the cross terms in the numerator, the proposed MRC approach cannot be used to solve for the above expectation. Since the second term is independent of P_t, we can only say that $I_{\text{OAF}} = \frac{1}{2}\log(P_t) + O(1)$ in high power regimes. In low powers, following a procedure similar to the NAF system, the achievable rate for both FG and CI systems can be approximated from (4.19) as

$$I_{\mathrm{OAF}} \approx \mathbb{E}\left[\frac{f_{\mathrm{OAF}}(\boldsymbol{h})}{2\ln(2)}\right] = \mathbb{E}\left[\frac{P_t q_1 \alpha_0 + O(P_t^2)}{2\ln(2)[1+O(P_t)]}\right] = \frac{P_t q_1 \phi_0}{2\ln(2)}. \tag{4.20}$$

As in the NAF scheme, $I_{\mathrm{OAF}}^{\mathrm{FG}} \geq I_{\mathrm{OAF}}^{\mathrm{CI}}$ at low powers and the CI system approaches the FG one from below.

Finally, consider the DT scheme in which the relay is not used for transmission with $z_2 = 0$. The unconditional rate is simply obtained by taking the expectation of (2.6) with respect to α_0. The rate in this case can be written in closed-form as

$$I_{\mathrm{DT}} = \frac{1}{2\ln(2)}\left[\mathscr{J}\left(\frac{1}{q_1 P_t \phi_0}\right) + \mathscr{J}\left(\frac{1}{q_2 P_t \phi_0}\right)\right], \tag{4.21}$$

which can be respectively approximated at high and low powers as

$$I_{\mathrm{DT}} \approx \log(P_t \phi_0 \sqrt{q_1 q_2}) - \frac{\gamma}{\ln(2)}, \quad \text{and} \quad I_{\mathrm{DT}} \approx \frac{P_t \phi_0 (q_1 + q_2)}{2\ln(2)}. \tag{4.22}$$

4.2.3 TW Systems

For the two-phase TWAF system using the amplification coefficients in (2.13), the conditional achievable rate in (2.14) can be written as

$$I_{i,\mathrm{TW}}|\boldsymbol{h} = \frac{1}{2}\log\left(1 + \frac{P_t^2 q_k z_2 \alpha_1 \alpha_2}{P_t(q_1 \Xi_1 + q_2 \Xi_2 + z_2 \alpha_i) + 1}\right), \tag{4.23}$$

where $\Xi_1 = \phi_1$ and $\Xi_2 = \phi_2$ for FG, and $\Xi_1 = \alpha_1$ and $\Xi_2 = \alpha_2$ for CI. Following the same approach as in the previous subsections, the expectation of (4.23) can be respectively approximated in high power regions for FG and CI systems as

$$\begin{aligned} I_{i,\mathrm{TW}}^{\mathrm{FG}} \approx& \frac{1}{2}\log(P_t q_k z_2) + \frac{\mathbb{E}[\ln(\alpha_1)] + \mathbb{E}[\ln(\alpha_2)] - \mathbb{E}[\ln(q_1\phi_1 + q_2\phi_2 + z_2\alpha_i)]}{2\ln(2)} \\ =& \frac{1}{2}\log(P_t) + \log\left(\sqrt{\frac{q_k z_2 \phi_1 \phi_2}{q_1\phi_1 + q_2\phi_2}}\right) - \frac{1}{2\ln(2)}\mathscr{J}\left(\frac{q_1\phi_1 + q_2\phi_2}{z_2\phi_i}\right) - \frac{\gamma}{\ln(2)}, \end{aligned} \tag{4.24}$$

and

$$I_{i,\mathrm{TW}}^{\mathrm{CI}} \approx \frac{1}{2}\log(P_t q_k z_2) + \frac{\mathbb{E}[\ln(\alpha_1)] + \mathbb{E}[\ln(\alpha_2)] - \mathbb{E}[\ln(q_1\alpha_1 + q_2\alpha_2 + z_2\alpha_i)]}{2\ln(2)}$$

$$= \frac{1}{2}\log(P_t) + \begin{cases} \log(\sqrt{z_2\phi_i}) - \frac{\gamma}{2\ln(2)} - \frac{1}{2\ln(2)}, & q_k\phi_k = (q_i + z_2)\phi_i \\ \frac{(q_i+z_2)\phi_i \log\left(\frac{q_i+z_2}{z_2 q_k \phi_k}\right) + q_k\phi_k \log(z_2\phi_i)}{2(q_k\phi_k - (q_i+z_2)\phi_i)} - \frac{\gamma}{2\ln(2)}, & q_k\phi_k \neq (q_i + z_2)\phi_i . \end{cases} \quad (4.25)$$

Using Jensen's inequality, it can be easily shown that $I_{i,\text{TW}}^{\text{CI}} \geq I_{i,\text{TW}}^{\text{FG}}$ at high powers. Thus, the CI system outperforms the FG one in high power regions.

In low power regions, the achievable rates for both FG and CI systems converge and can simply be approximated from (4.23) as

$$I_{i,\text{TW}} \approx \frac{P_t^2 q_k z_2 \phi_1 \phi_2}{2\ln(2)}. \quad (4.26)$$

In this case, $f_i^{\text{FG}}(\boldsymbol{h}) - f_{i,2\text{W}}^{\text{CI}}(\boldsymbol{h})$ is neither convex nor concave and thus one cannot say which of the two coefficients is dominant at low powers.

4.2.4 Remarks

First, observe from (4.9), (4.10), (4.15), (4.16), (4.24) and (4.25) that the derived high power approximations are easy to evaluate as they involve only the exponential integral. In general, these high power approximations can be written as

$$I \approx m \cdot \log(P_t) + \mathcal{G}_{\text{high}}(\boldsymbol{q}, \boldsymbol{\Phi}), \quad (4.27)$$

where $m \in \{1/2, 1\}$ is the multiplexing gain and $\mathcal{G}_{\text{high}}(\cdot,\cdot)$ is the power gain. As shown in the previous subsections, the multiplexing gain depends only on the transmission protocol, whereas the power gain also depends on the power allocation $\boldsymbol{q}$, the network configuration $\boldsymbol{\Phi}$, and the selected amplification coefficient (FG or CI). As shall be shown in Sect. 4.5, the derived bounds are tight in medium to high power regions. The impact of the power gain $\mathcal{G}_{\text{high}}(\cdot,\cdot)$ can thus be significant in these power regimes, as will be illustrated shortly.

In low power regions, it can be seen from previous subsections that the derived approximations can be written as

$$I \approx P_t^n \cdot \mathcal{G}_{\text{low}}(\boldsymbol{q}, \boldsymbol{\Phi}). \quad (4.28)$$

Similar to (4.27), $n \in \{1, 2\}$ only depends on the transmission protocol, whereas $\mathcal{G}_{\text{low}}(\cdot,\cdot)$ also depends on the power allocation, network configuration and amplification coefficient.

For ease of reference, Tables 4.1 and 4.2 summarize the approximations derived in this section in terms of n, m, $\mathcal{G}_{\text{high}}(\cdot,\cdot)$ and $\mathcal{G}_{\text{low}}(\cdot,\cdot)$. In these tables, a star ($\star$) has been placed next to the dominant amplification coefficient when known. As shall be illustrated in Sect. 4.5, the approximations in Tables 4.1 and 4.2 are tight in

Table 4.1 High power approximations for different AF protocols

Protocol	Type	m	$\mathcal{G}_{\text{high}}(\boldsymbol{q}, \boldsymbol{\Phi})$
DH	CI*	1/2	$\begin{cases} \log(\sqrt{q_1\phi_1}) - \frac{\gamma}{2\ln(2)} - \frac{1}{2\ln(2)}, & q_1\phi_1 = z_2\phi_2 \\ \frac{q_1\phi_1 \log(z_2\phi_2) - z_2\phi_2 \log(q_1\phi_1)}{2(q_1\phi_1 - z_2\phi_2)} - \frac{\gamma}{2\ln(2)}, & q_1\phi_1 \neq z_2\phi_2 \end{cases}$
	FG	1/2	$\log(\sqrt{z_2\phi_2}) - \frac{1}{2\ln(2)} \mathcal{J}\left(\frac{q_1\phi_1}{z_2\phi_2}\right) - \frac{\gamma}{\ln(2)}$
NAF	CI	1	$\begin{cases} \log(\phi_0\sqrt{q_1 q_2}) - \frac{\gamma}{\ln(2)} - \frac{1}{2\ln(2)}, & q_1\phi_1 = z_2\phi_2 \\ \log(\phi_0 q_1\sqrt{q_2\phi_1}) - \frac{q_1\phi_1 \ln(q_1\phi_1) - z_2\phi_2 \ln(z_2\phi_2)}{2\ln(2)(q_1\phi_1 - z_2\phi_2)} - \frac{\gamma}{\ln(2)}, & q_1\phi_1 \neq z_2\phi_2 \end{cases}$
	FG*	1	$\log(\phi_0\sqrt{q_1 q_2}) - \frac{1}{2\ln(2)} \mathcal{J}\left(\frac{q_1\phi_1}{z_2\phi_2}\right) - \frac{\gamma}{\ln(2)}$
OAF	CI/FG	1/2	$O(1)$
DT	–	1	$\log(\phi_0\sqrt{q_1 q_2}) - [\gamma/\ln(2)]$
TW	CI*	1/2	$\begin{cases} \log(\sqrt{z_2\phi_i}) - \frac{\gamma}{2\ln(2)} - \frac{1}{2\ln(2)}, & q_k\phi_k = (q_i + z_2)\phi_i \\ \frac{(q_i+z_2)\phi_i \log\left(\frac{q_i+z_2}{z_2 q_k\phi_k}\right) + q_k\phi_k \log(z_2\phi_i)}{2(q_k\phi_k - (q_i+z_2)\phi_i)} - \frac{\gamma}{2\ln(2)}, & q_k\phi_k \neq (q_i + z_2)\phi_i \end{cases}$
	FG	1/2	$\log\left(\sqrt{\frac{q_k z_2\phi_1\phi_2}{q_1\phi_1 + q_2\phi_2}}\right) - \frac{1}{2\ln(2)} \mathcal{J}\left(\frac{q_1\phi_1 + q_2\phi_2}{z_2\phi_i}\right) - \frac{\gamma}{\ln(2)}$

Table 4.2 Low power approximations for different AF protocols

Protocol	Type	n	$\mathcal{G}_{\text{low}}(\boldsymbol{q}, \boldsymbol{\Phi})$
DH	CI / FG*	2	$\frac{q_1 z_2\phi_1\phi_2}{2\ln(2)}$
NAF	CI / FG*	1	$\frac{(q_1+q_2)\phi_0}{2\ln(2)}$
OAF	CI / FG*	1	$\frac{q_1\phi_0}{2\ln(2)}$
DT	–	1	$\frac{\phi_0(q_1+q_2)}{2\ln(2)}$
TW	CI / FG	2	$\frac{q_k z_2\phi_1\phi_2}{2\ln(2)}$

their respective power regions. The problems in (4.1) and (4.2) can thus be solved by maximizing the approximations in these two tables without the need of lengthy Monte Carlo simulations.

4.3 Optimal Power Allocation

In this section, optimal power allocation schemes that maximize the achievable rate in (4.1) or the sum rate in (4.2) are investigated. As before, we focus on high and low power regions.

4.3.1 DHAF Systems

The capacity of the DH system, denoted as C_{DH}, using either the FG or the CI coefficient in (2.4) can be obtained by maximizing the expectation of (4.7) over the channel gains. Since (4.7) is not a function of q_2, the feasible region in (4.1) reduces to $q_1 + z_2 \leq q_t$. By taking the first derivatives, it can be easily shown that (4.7) is strictly increasing with q_1 and z_2. Hence, the power constraint in (4.1) must be tight $q_1 + z_2 = q_t$. Furthermore, the second derivative of (4.7) over the line segment $z_2 = q_t - q_1$ $(0 \leq q_1 \leq q_t)$ can be shown to be negative. The maximization of the achievable rate is thus a concave optimization problem for DH systems. The optimal power allocation $\boldsymbol{q}_{\text{DH}} = [q_{1,\text{DH}}, z_{2,\text{DH}}] = [q_{1,\text{DH}}, q_t - q_{1,\text{DH}}]$ is then unique and can be easily obtained by finding a stationary point in the rate.

At high enough transmission powers, the optimal power allocation that achieves the capacity in (4.1) with CI or FG can be obtained by maximizing the approximations in (4.9) or (4.10) over the line $z_2 = q_t - q_1$. Since (4.9) and (4.10) are non-linear, finding a closed-form expression for the unique stationary point $q_{1,\text{DH}}$ appears difficult. However, a suboptimal power allocation in high power regions can be derived as follows. Using (4.11) and the fact that $\mathscr{J}(x) < \ln(1 + [1/x])$ [1], the achievable rates in (4.9) and (4.10) can be lower bounded as

$$I_{\text{DH}}^{\text{CI}} \geq I_{\text{DH}}^{\text{FG}} \geq \frac{\log(P_t)}{2} + \frac{1}{2}\log\left(\frac{q_1 z_2 \phi_1 \phi_2}{q_1 \phi_1 + z_2 \phi_2}\right) - \frac{\gamma}{\ln(2)}. \tag{4.29}$$

The maximization of the bound in (4.29) can also be shown to be a concave problem. By setting $z_2 = q_t - q_1$ and equating the derivative of (4.29) to zero, the suboptimal power allocation $\boldsymbol{q}_{\text{DH}}^{\text{sub}} = [q_{1,\text{DH}}^{\text{sub}}, q_t - q_{1,\text{DH}}^{\text{sub}}]$ at high powers can be calculated as

$$q_{1,\text{DH}}^{\text{sub}} = \begin{cases} \frac{q_t(\phi_2 - \sqrt{\phi_1 \phi_2})}{\phi_2 - \phi_1}, & \phi_1 \neq \phi_2 \\ q_t/2, & \phi_1 = \phi_2. \end{cases} \tag{4.30}$$

As shall be illustrated in Sect. 4.5, the system using the suboptimal power allocation in (4.30) performs closely to the one using the optimal power allocation that achieves the capacity in (4.1) for different network configurations.

In low power regions, it is easy to see from (4.12) and the arithmetic/geometric mean inequality that uniform power allocation $q_{1,\text{DH}} = z_{2,\text{DH}} = q_t/2$ is optimal for CI and FG systems. This holds regardless of the value of $\boldsymbol{\Phi}$.

4.3.2 Cooperative Systems

For the cooperative systems, one must find the power allocation $\boldsymbol{q}$ that maximizes the expectation of (4.14) as in (4.1). As observed in [5] and Chap. 3, the maximization in (4.1) for cooperative systems is in general not a concave problem. Given that a diagonal covariance matrix $\boldsymbol{Q}$ is optimal [5] for fading channels, it is easy to show following a similar approach as in Chap. 3 that the capacity can be written as $C = \max\{C_{\text{DT}}, C_{\text{OAF}}, C_{\text{NAF}}\}$, where C_{DT}, C_{OAF} and C_{NAF} are the capacities of the DT, OAF and NAF protocols, respectively.

First, it can be shown that the mutual information of the DT system in (4.21) is maximized when $q_{1,\text{DT}} = q_{2,\text{DT}} = q_t/2$ for any value of P_t. The capacity of the DT scheme is then given as $C_{\text{DT}} = [1/\ln(2)] \cdot \mathscr{J}(2/[P_t\phi_0 q_t])$. At high powers, the capacity of the NAF channel can be calculated by maximizing the approximations in (4.15) and (4.16) for the FG and CI system, respectively. Observe from (4.15) that $I_{\text{NAF}}^{\text{FG}}$ is a decreasing function of z_2 since $\mathscr{J}(x) > 0$ is a decreasing function of $x > 0$. This means that using the relay in fact reduces the rate in high power regions. From this and (4.17), $I_{\text{DT}} \geq I_{\text{NAF}}^{\text{FG}} \geq I_{\text{NAF}}^{\text{CI}}$, where the equality is achieved when $z_{2,\text{NAF}} = 0$. Given that I_{DT} in (4.21) is maximized when $q_{1,\text{DT}} = q_{2,\text{DT}} = q_t/2$, we have the following inequality

$$C_{\text{DT}} \geq I_{\text{DT}} \geq I_{\text{NAF}}^{\text{FG}} \geq I_{\text{NAF}}^{\text{CI}}, \tag{4.31}$$

where the equality is achieved when $q_{1,\text{NAF}} = q_{2,\text{NAF}} = q_t/2$ and $z_{2,\text{NAF}} = 0$. The optimal power allocation for both FG and CI NAF systems is thus $\boldsymbol{q}_{\text{NAF}} = [q_t/2, q_t/2, 0]$. The capacity of the NAF system is then $C_{\text{NAF}} = C_{\text{DT}}$ and is achieved when the relay is inactive. This observation is similar to that shown in [5] for symmetric $\phi_0 = \phi_1 = \phi_2$ NAF channels using the FG coefficient. Furthermore, observe from Table 4.1 that the rate of the OAF system only grows as $\frac{1}{2}\log(P_t)$ rather than logarithmically with P_t as the rate of the DT scheme. Hence, $C_{\text{DT}} = C_{\text{NAF}} > C_{\text{OAF}}$ at sufficiently high powers. Therefore, the capacity in (4.1) for cooperative systems operating in high power regions is achieved by the DT scheme.

In low powers, it can be seen from Table 4.2 that the achievable rate of the NAF and OAF systems is independent of the power allocated to the relay z_2. As a result, the maximum rate is achieved when $z_2 = 0$. Therefore, at sufficiently low powers, the relay is inactive and the optimal power allocations are $\boldsymbol{q}_{\text{NAF}} = [q_t/2, q_t/2, 0]$ and $\boldsymbol{q}_{\text{OAF}} = [q_t, 0, 0]$. Given that $I_{\text{DT}}|\boldsymbol{h} > I_{\text{OAF}}|\boldsymbol{h}$ when $q_{1,\text{DT}} = q_{2,\text{DT}} = q_t/2$ and $\boldsymbol{q}_{\text{OAF}} = [q_t, 0, 0]$, $C_{\text{DT}} = C_{\text{NAF}} > C_{\text{OAF}}$ and the capacity in (4.1) is then also achieved by the DT scheme in low power regimes.

4.3.3 TW Systems

For TWAF systems, our objective is to maximize the sum rate as in (4.2). For the TW system operating in high power regions, the approximations in (4.24) and (4.25) must be used. Similar to DH systems, this is a non-linear problem and numerical methods are required. A suboptimal power allocation can, however, be obtained as in (4.30). Similar to (4.29), the sum rate of the FG and CI systems can be lower bounded as

$$I_{\text{sum}}^{\text{TW, CI}} \geq I_{\text{sum}}^{\text{TW, FG}} \geq \frac{1}{2}\log\left(\frac{P_t^2 q_1 q_2 z_2^2 \phi_1^2 \phi_2^2}{(q_1\phi_1 + q_2\phi_2 + z_2\phi_1)(q_1\phi_1 + q_2\phi_2 + z_2\phi_2)}\right) - \frac{2\gamma}{\ln(2)}. \tag{4.32}$$

Although the above bound is not concave, a global maximizer can still be found using the approach in [11]. From (4.30) and [11], it can be shown that the power allocation that maximizes (4.32), $\boldsymbol{q}_{\text{TW}}^{\text{sub}} = [q_{1,\text{TW}}^{\text{sub}}, q_{2,\text{TW}}^{\text{sub}}, z_{2,\text{TW}}^{\text{sub}}]$, is given by

$$\boldsymbol{q}_{\text{TW}}^{\text{sub}} = \begin{cases} \left[\frac{q_t(\phi_2 - \sqrt{\phi_1\phi_2})}{2(\phi_2-\phi_1)}, \frac{q_t(\sqrt{\phi_1\phi_2}-\phi_1)}{2(\phi_2-\phi_1)}, q_t/2\right], & \phi_1 \neq \phi_2 \\ [q_t/4, q_t/4, q_t/2], & \phi_1 = \phi_2. \end{cases} \tag{4.33}$$

As will be shown in Sect. 4.5, the power allocation in (4.33) approaches closely to the optimal one and provides great advantages over the schemes previously proposed in [10, 21].

In low power regions, the sum rate of the TW system using either CI or FG can be approximated from (4.26) as

$$I_{\text{sum}}^{\text{TW}} \approx \frac{P_t^2 z_2 (q_1 + q_2)\phi_1\phi_2}{2\ln(2)}. \tag{4.34}$$

From the inequality of the geometric and arithmetic means, it is not difficult to see that the sum rate in (4.34) is maximized when 50 % of the power is given to the relay, while the other 50 % is equally distributed between the two sources. The optimal power allocation and the solution to (4.2) at low powers is then $\boldsymbol{q}_{\text{TW}} = [q_t/4, q_t/4, q_t/2]$.

4.4 Closed-Form Solution for DHAF System with CI Coefficient

For the specific case of the DH system with the CI coefficient, closed-form expressions of the unconditional achievable rate can be obtained using the MRC approach in Proposition 4.1. Specifically, factoring $f_{1,\text{DH}}(\boldsymbol{h}) + f_{2,\text{DH}}(\boldsymbol{h}) = (1 + P_t q_1 \alpha_1)(1 + P_t z_2 \alpha_2)$ in (4.7), the achievable rate of the CI system can be expressed as

$$\begin{aligned} I_{\text{DH}}^{\text{CI}} &= 0.5 \cdot \mathbb{E}[\log(1 + P_t q_1 \alpha_1)] + 0.5 \cdot \mathbb{E}[\log(1 + P_t z_2 \alpha_2)] \\ &\quad - 0.5 \cdot \mathbb{E}[\log(1 + P_t q_1 \alpha_1 + P_t z_2 \alpha_2)]. \end{aligned} \tag{4.35}$$

From (4.5) in Proposition 4.1, the expectations in (4.35) can be easily solved as

$$I_{\text{DH}}^{\text{CI}} = \begin{cases} \frac{1}{2\ln(2)}\left[\left(1+\frac{1}{P_t q_1 \phi_1}\right)\mathscr{J}\left(\frac{1}{P_t q_1 \phi_1}\right)-1\right], & q_1\phi_1 = z_2\phi_2 \\ \frac{q_1\phi_1 \mathscr{J}\left(\frac{1}{P_t z_2 \phi_2}\right) - z_2\phi_2 \mathscr{J}\left(\frac{1}{P_t q_1 \phi_1}\right)}{2\ln(2)(q_1\phi_1 - z_2\phi_2)}, & q_1\phi_1 \neq z_2\phi_2. \end{cases} \tag{4.36}$$

Observe that (4.36) is in closed-form and holds for any value of P_t.

Given that a closed-form expression is available, (4.36) can also be used to derive optimal power allocations as in Sect. 4.3. Recall from Sect. 4.3.1 that the mutual information of the DH system is concave along the line $z_2 = q_t - q_1$. The optimal allocation can then simply be obtained by finding the unique stationary point of (4.36) along this line. With a slight abuse of notation, let $I([q_1, z_2]) = I_{\text{DH}}^{\text{CI}}$ in (4.36). The derivative of $I([q_1, q_t - q_1])$ can be written when $q_1\phi_1 \neq (q_t - q_1)\phi_2$ as

$$\frac{dI([q_1, q_t - q_1])}{dq_1} = \frac{-Aq_1^4 - Bq_1^2 + Eq_1 + Hq_1^5 + Fq_1^3 - G}{P_t q_1^2 \phi_2 \phi_1 (q_t - q_1)^2 [(q_t - q_1)\phi_2 - q_1\phi_1]^2}, \tag{4.37}$$

where we have used the fact that $d\mathscr{J}(x)/dx = \mathscr{J}(x) - [1/x]$ [1]. In (4.37), $A = \mathscr{J}_2\phi_2^2(\phi_2 + \phi_1 - P_t q_t \phi_1^2) + \mathscr{J}_1\phi_1^2(P_t q_t \phi_2^2 - \phi_1 - \phi_2) + P_t q_t \phi_2\phi_1(\phi_1^2 + 4\phi_2^2 + 5\phi_2\phi_1)$, $B = q_t^2\phi_2^2[P_t q_t \phi_1(\mathscr{J}_1\phi_1 + \phi_1 + 4\phi_2) + (6\phi_2 + 3\phi_1 - P_t q_t \phi_1^2)\mathscr{J}_2]$, $H = P_t\phi_2\phi_1(\phi_2 + \phi_1)^2$, $F = q_t\phi_2[\mathscr{J}_1\phi_1^2(2P_t q_t \phi_2 - 1) + 2P_t q_t \phi_2\phi_1(2\phi_1 + 3\phi_2) + \mathscr{J}_2\phi_2(4\phi_2 + 3\phi_1 - 2P_t q_t \phi_1^2)]$, $E = q_t^3\phi_2^2(4\mathscr{J}_2\phi_2 + \mathscr{J}_2\phi_1 + P_t q_t \phi_2\phi_1)$, and $G = \mathscr{J}_2 q_t^4 \phi_2^3$, with $\mathscr{J}_1 = \mathscr{J}(1/[P_t(q_t - q_1)\phi_2])$ and $\mathscr{J}_2 = \mathscr{J}(1/[P_t q_1 \phi_1])$. When $q_1\phi_1 = (q_t - q_1)\phi_2$, $q_1 = (q_t\phi_2)/(\phi_2 + \phi_1)$ and (4.37) can be simplified to

$$\frac{d}{dq_1} I\left(\left[\frac{q_t\phi_2}{\phi_2 + \phi_1}, \frac{q_t\phi_1}{\phi_2 + \phi_1}\right]\right) = \frac{(\phi_2^2 - \phi_1^2)K}{2q_t^3\phi_2^3\phi_1^3 P_t^2}, \tag{4.38}$$

where $K = [\phi_1 + \phi_2][(\phi_1 + \phi_2)\mathscr{J}_3 + 2P_t q_t \phi_2\phi_1\mathscr{J}_3 - P_t q_t \phi_2\phi_1] - [P_t q_t \phi_2\phi_1]^2$ and $\mathscr{J}_3 = \mathscr{J}([\phi_2 + \phi_1]/[P_t q_t \phi_2\phi_1])$. The optimal power allocation can then be obtained by finding the point $0 \leq q_1 \leq q_t$ such that (4.37) is equal to zero. Given that (4.37) is highly non-linear, finding a closed-form expression for the stationary point $q_{1,\text{DH}}^{\text{CI}}$ is not straightforward. From the concavity of (4.36), the optimal power allocation can be obtained numerically by performing bisection on (4.37) for $0 \leq q_1 \leq q_t$.

Observe from (4.38) that $I([q_1, q_t - q_1])$ has a stationary point when $\phi_2 = \phi_1$ at $q_1 = (q_t\phi_2)/(\phi_2 + \phi_1)$, i.e., when the channel is symmetric. Since $I([q_1, q_t - q_1])$ is concave, $q_{1,\text{DH}}^{\text{CI}} = z_{2,\text{DH}}^{\text{CI}} = q_t/2$ must be the global maximizer and thus uniform power allocation is optimal for the symmetric network.

4.5 Illustrative Examples

In this section, simulation results are provided to confirm the analysis in this chapter. For all simulations, we adopt the linear network model similar to Chap. 3, where the relay is in the line between the source and destination, the $S_1 - S_2$ distance is normalized to 1, and the $S_1 - R$ and $R - S_2$ distances are d and $(1-d)$, respectively, $(0 \leq d \leq 1)$. In this model, $\phi_0 = 1$, $\phi_1 = 1/d^\nu$ and $\phi_2 = 1/(1-d)^\nu$, where $\nu = 3$ is the pathloss exponent. For convenience, the results in this section arc plotted against P_t/N_0 and $q_t = 2$. In addition, three types of power allocation schemes are considered: (1) uniform power allocation, i.e., $\boldsymbol{q} = [q_t/3, q_t/3, q_t/3]$ for NAF/TW, or $\boldsymbol{q} = [q_t/2, 0, q_t/2]$ for DH/OAF; (2) the optimal power allocation that maximizes the achievable rate or sum rate as in (4.1) or (4.2) obtained using exhaustive search; and (3) the power allocation schemes proposed in Sect. 4.3.

4.5.1 Tightness of the Proposed Approximations

To verify the high power approximations derived in Table 4.1, Fig. 4.1 shows the single-user achievable rates of the DH, TW and NAF systems using uniform power allocation at $d = 0.5$. The rate of the NAF system has been pre-multiplied by a factor of $1/2$ for clarity of the figure. Observe from Fig. 4.1 that all approximations coincide with the Monte Carlo simulations in medium to high powers. Although all rates in Fig. 4.1 have the same slope, i.e., the same multiplexing gain, it can be seen that the power gains vary widely among the considered protocols. It can also be observed from Fig. 4.1 that at sufficiently high powers, the CI technique is superior for DH and TW systems, while the FG one is better for the NAF protocol, which confirms the analysis in Sect. 4.2.

Figure 4.2 shows the achievable rates of the systems using uniform power allocation along with the low power approximations in Table 4.2. Observe again from Fig. 4.2 that the simple approximations in Table 4.2 match with the simulation results at a sufficiently low powers for all systems. As discussed in Sect. 4.2, the OW

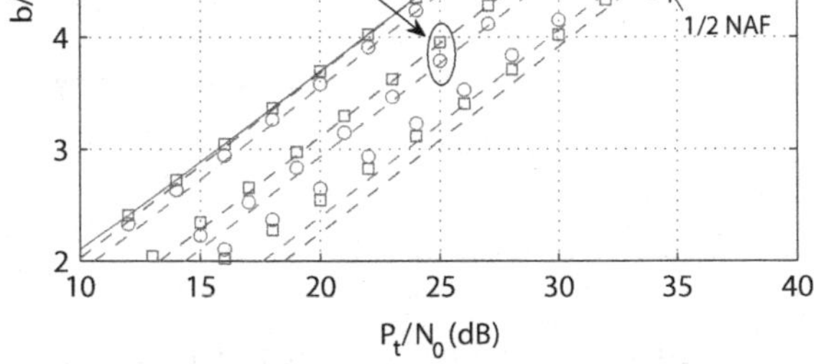

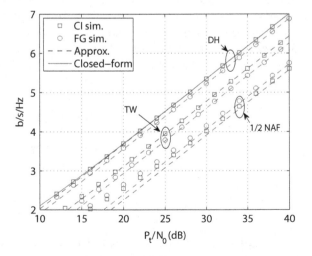

Fig. 4.1 Achievable rates and high power approximations with uniform power allocation ($d = 0.5$)

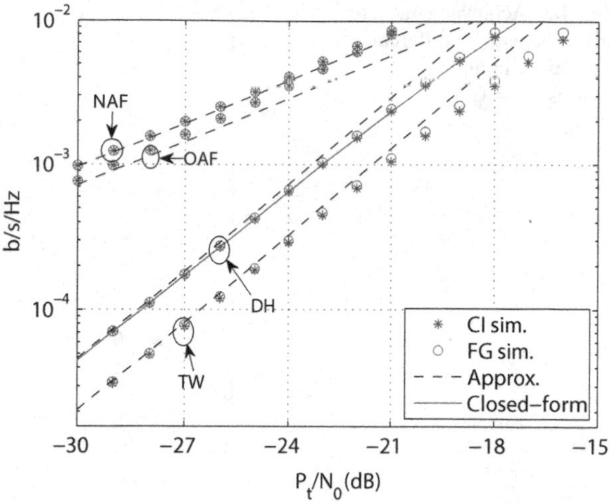

Fig. 4.2 Achievable rates and low power approximations with uniform power allocation ($d = 0.5$)

FG systems are slightly better than the CI ones in low power regimes. However, it can be seen in Fig. 4.2 that the two amplification coefficients present an almost identical performance as P_t/N_0 decreases.

Besides the derived approximations, the closed-form expression of the achievable rate for the DH CI system in (4.36) is also shown in Figs. 4.1 and 4.2. Note from these figures that the expression derived in Sect. 4.4 matches with the Monte Carlo simulations for all values of P_t/N_0, as expected.

4.5.2 Optimal Power Allocation

To quantify the advantage of the power allocation schemes proposed in Sect. 4.3, this subsection compares the achievable rates of the considered systems using different power allocation strategies. As an example, we set $d = 0.3$. We shall use solid lines to represent the CI technique and dotted ones to represent the FG technique.

Figure 4.3 shows the achievable rates of the DH system in high power regimes. Besides the proposed power allocation in (4.30), we also consider the uniform and the optimal allocation that achieves the capacity in (4.1), which was obtained using exhaustive search. For the CI system, we further consider the allocation obtained by applying bisection to the closed-form derivative in (4.37). Figure 4.3 indicates that the systems using the proposed allocation schemes outperform those using uniform allocation. Specifically, asymptotic gains of 0.8 dB and 1.2 dB can be observed for the CI and FG systems, respectively. More importantly, the proposed power allocations present a negligible loss over the optimal one for both amplification techniques. The rate difference between (4.30) and the bisection method in (4.37) is also negligible for the CI system.

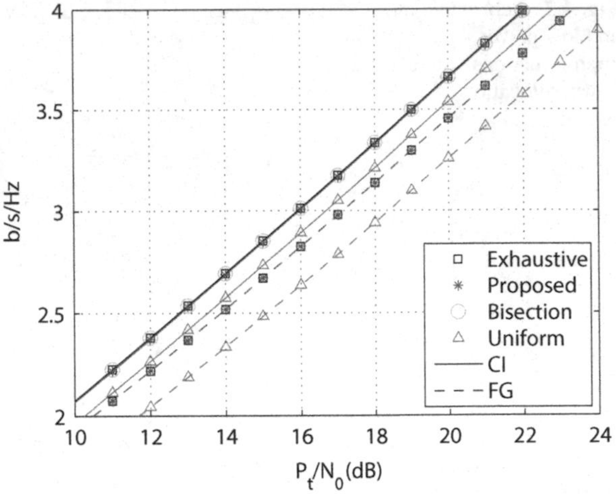

Fig. 4.3 Achievable rates of DH systems with different allocations in high power regions ($d = 0.3$)

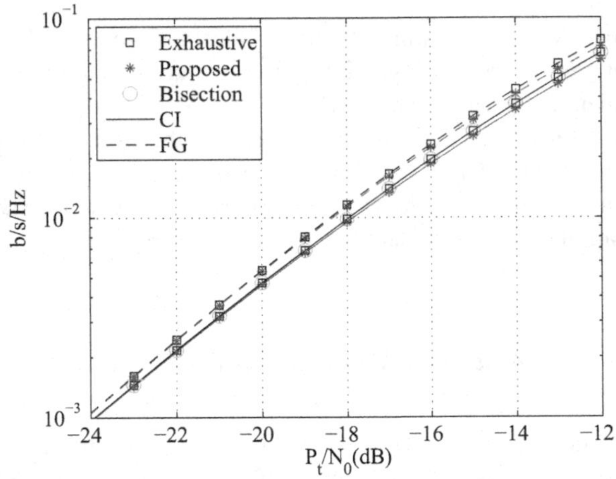

Fig. 4.4 Achievable rates of DH systems with different allocations in low power regions ($d = 0.3$)

The achievable rate of the DH system in low power regions is shown in Fig. 4.4. Recall from Sect. 4.3 that uniform power allocation $q_{1,\text{DH}} = z_{2,\text{DH}} = 1$ is asymptotically optimal in low powers. Thus, only the uniform, optimal (exhaustive) and bisection (for CI) power allocations are examined in Fig. 4.4. As expected, the uniform power allocation scheme presents a negligible loss over the optimal one at sufficiently low powers.

The single-user achievable rates of the NAF and OAF systems are shown in Figs. 4.5 and 4.6 in high and low power regions, respectively. Recall from Sect. 4.3 that for the NAF system, $\boldsymbol{q}_{\text{NAF}} = [1, 1, 0]$ is optimal in high and low power scenarios, i.e., DT is optimal. As such, it can be seen from these figures that the

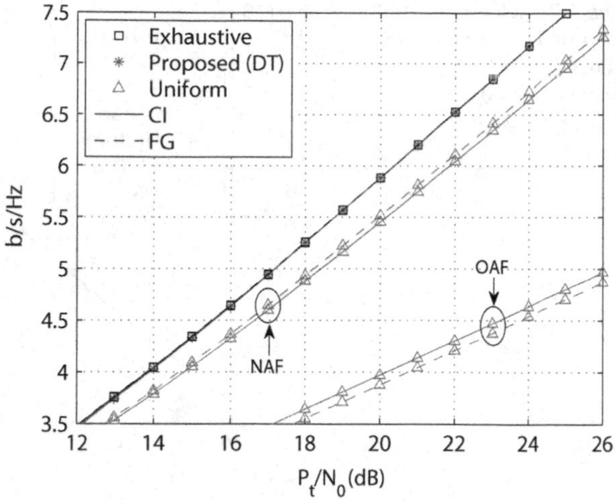

Fig. 4.5 Achievable rates of NAF/OAF with different allocations in high power regions ($d = 0.3$)

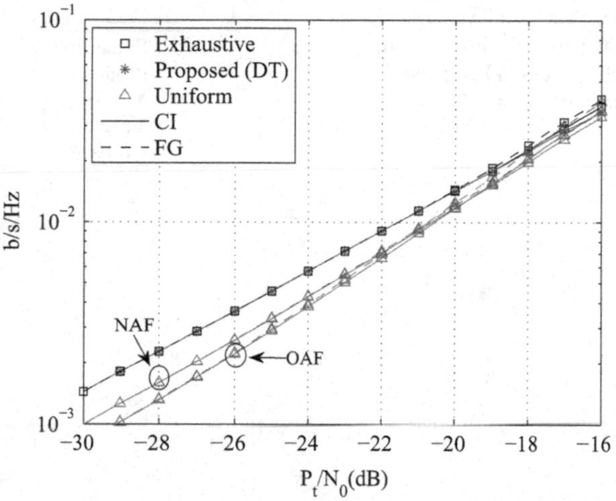

Fig. 4.6 Achievable rates of NAF/OAF with different allocations in low power regions ($d = 0.3$)

NAF systems using the proposed and the exhaustive power allocation present an indistinguishable performance and clearly outperform the OAF/NAF schemes with uniform power allocation.

The sum rate performance of the TW system with different power allocation strategies is shown in Fig. 4.7 for high power regions. Besides the uniform and optimal allocations, the schemes proposed in [10, 21] for the CI system are also considered in Fig. 4.7. Observe from Fig. 4.7 that in high power regions, the allocation proposed in (4.33) performs closely to the optimal one and provides significant gains over the other schemes. In particular, the CI system using the proposed power allocation presents a 1.3 dB gain over the system with uniform power allocation and a 0.5 dB gain over the systems using the schemes previously

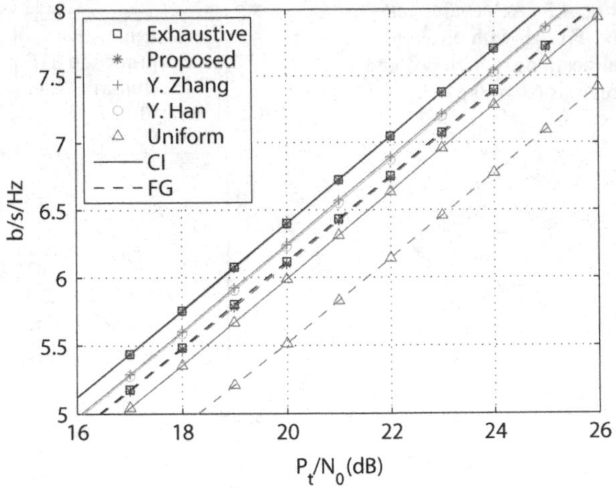

Fig. 4.7 Achievable sum rates of TW with different allocations in high power regions ($d = 0.3$)

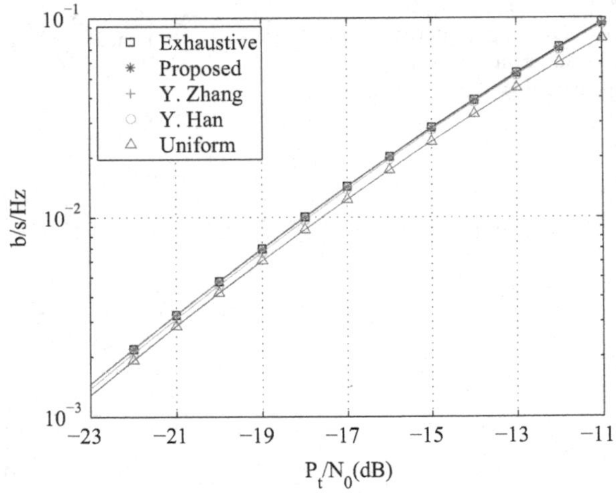

Fig. 4.8 Achievable sum rates of TW with different allocations in low power regions ($d = 0.3$)

proposed in [10, 21]. Using the proposed power allocation solution, a significant gain of 1.9 dB can also be achieved over the uniform power allocation scheme in FG systems.

Figure 4.8 shows the sum rates of the TW systems using the CI technique in low power regions. The rates of the FG systems are omitted in Fig. 4.8 for clarity of the figure. Observe from Fig. 4.8 that as P_t decreases, the performance of the system using the proposed asymptotic power allocation $\boldsymbol{q}_{\mathrm{TW}} = [1/2, 1/2, 1]$ closely approaches that of the system using the power allocation from exhaustive search. These two power allocations present a 0.5 dB gain over the uniform scheme. The allocations from [10, 21] also present a comparable performance. Given that the rates of the CI and FG techniques converge at low powers, a similar behavior can be obtained for the FG systems.

4.6 Concluding Remarks

In this chapter, achievable rates and power allocation schemes were investigated for OW and TW single-relay AF systems over Rayleigh fading channels. Tight high and low power approximations were first derived for several relaying protocols. The achievable rates of the systems using the CI and FG coefficients were then compared. From the derived approximations, asymptotic power allocation strategies were also proposed for the protocols of interest. Finally, simulation results were provided to verify the accuracy of the approximations and quantify the gain of the proposed allocation schemes.

References

1. Abramowitz, M., Stegun, I.A.: Handbook of Mathematical Functions. Dover Publications, New York (1965)
2. Beaulieu, N., Soliman, S.: Exact analysis of multihop amplify-and-forward relaying systems over general fading links. IEEE Trans. Commun. **60**(8), 2123–2134 (2012). Doi:10.1109/TCOMM.2012.051412.110343
3. Beaulieu, N., Farhadi, G., Chen, Y.: A precise approximation for performance evaluation of amplify-and-forward multihop relaying systems. IEEE Trans. Wireless Commun. **10**(12), 3985–3989 (2011). Doi:10.1109/TWC.2011.101811.101133
4. Di Renzo, M., Graziosi, F., Santucci, F.: A comprehensive framework for performance analysis of dual-hop cooperative wireless systems with fixed-gain relays over generalized fading channels. IEEE Trans. Wireless Commun. **8**(10), 5060–5074 (2009). Doi:10.1109/TWC.2009.080318
5. Ding, Y., Zhang, J.K., Wong, K.M.: Ergodic channel capacities for the amplify-and-forward half-duplex cooperative systems. IEEE Trans. Inf. Theory **55**(2), 713–730 (2009). Doi:10.1109/TIT.2008.2009822
6. Duong, T., Hoang, L.N., Bao, V.: On the performance of two-way amplify-and-forward relay networks. IEICE Trans. Commun. **E92-B**(12), 3957–3959 (2009)
7. Farhadi, G., Beaulieu, N.: On the ergodic capacity of wireless relaying systems over Rayleigh fading channels. IEEE Trans. Wireless Commun. **7**(11), 4462–4467 (2008). Doi:10.1109/T-WC.2008.070737
8. Farhadi, G., Beaulieu, N.: On the ergodic capacity of multi-hop wireless relaying systems. IEEE Trans. Wireless Commun. **8**(5), 2286–2291 (2009). Doi:10.1109/TWC.2009.080818
9. Firag, A., Smith, P., McKay, M.: Capacity analysis for MIMO two-hop amplify-and-forward relaying systems with the source to destination link. In: Proceedings of the IEEE International Conference Communications, pp. 1–6 (2009). Doi:10.1109/ICC.2009.5198803
10. Han, Y., Ting, S.H., Ho, C.K., Chin, W.H.: Performance bounds for two-way amplify-and-forward relaying. IEEE Trans. Wireless Commun. **8**(1), 432–439 (2009). Doi:10.1109/T-WC.2009.080316
11. Jang, Y.U., Jeong, E.R., Lee, Y.: A two-step approach to power allocation for OFDM signals over two-way amplify-and-forward relay. IEEE Trans. Signal Process. **58**(4), 2426–2430 (2010). Doi:10.1109/TSP.2010.2040415
12. Jiménez-Rodríguez, L., Tran, N.H., Le-Ngoc, T.: Achievable rate and power allocation for single-relay AF systems over Rayleigh fading channels at high and low SNRs. IEEE Trans. Veh. Technol. **63**(4), 1726–1739 (2014). Doi:10.1109/TVT.2013.2287997

13. Jin, S., McKay, M., Zhong, C., Wong, K.K.: Ergodic capacity analysis of amplify-and-forward MIMO dual-hop systems. IEEE Trans. Inf. Theory **56**(5), 2204–2224 (2010). Doi:10.1109/TIT.2010.2043765
14. Shao, J., Alouini, M.S., Goldsmith, A.: Impact of fading correlation and unequal branch gains on the capacity of diversity systems. In: Proceedings of IEEE Vehicular Technology Conference, vol. 3, pp. 2159–2163 (1999). Doi:10.1109/VETEC.1999.778436
15. Simon, M., Alouini, M.S.: Digital Communication over Fading Channels. Wiley-IEEE Press, Hoboken (2004)
16. Soliman, S., Beaulieu, N.: Exact analysis of dual-hop AF maximum end-to-end SNR relay selection. IEEE Trans. Commun. **60**(8), 2135–2145 (2012). Doi:10.1109/TCOMM.2012.061112.11-551
17. Waqar, O., Ghogho, M., McLernon, D.: Tight bounds for ergodic capacity of dual-hop fixed-gain relay networks under Rayleigh fading. IEEE Commun. Lett. **15**(4), 413–415 (2011). Doi:10.1109/LCOMM.2011.022411.102027
18. Waqar, O., McLernon, D., Ghogho, M.: Exact evaluation of ergodic capacity for multihop variable-gain relay networks: A unified framework for generalized fading channels. IEEE Trans. Veh. Technol. **59**(8), 4181–4187 (2010). Doi:10.1109/TVT.2010.2063047
19. Xia, M., Xing, C., Wu, Y.C., Aïssa, S.: Exact performance analysis of dual-hop semi-blind AF relaying over arbitrary Nakagami-m fading channels. IEEE Trans. Wireless Commun. **10**(10), 3449–3459 (2011). Doi:10.1109/TWC.2011.080311.102267
20. Yang, J., Fan, P., Duong, T., Lei, X.: Exact performance of two-way AF relaying in Nakagami-m fading environment. IEEE Trans. Wireless Commun. **10**(3), 980–987 (2011). Doi:10.1109/TWC.2011.011111.101141
21. Zhang, Y., Ma, Y., Tafazolli, R.: Power allocation for bidirectional AF relaying over Rayleigh fading channels. IEEE Commun. Lett. **14**(2), 145–147 (2010). Doi:10.1109/LCOMM.2010.02.092227
22. Zhong, C., Matthaiou, M., Karagiannidis, G., Ratnarajah, T.: Generic ergodic capacity bounds for fixed-gain AF dual-hop relaying systems. IEEE Trans. Veh. Technol. **60**(8), 3814–3824 (2011). Doi:10.1109/TVT.2011.2167362

Chapter 5
Half-Duplex AF Relaying: Adaptation Policies

In Chap. 4, we investigated the capacity of several AF protocols over fading channels. Similar to previous works in the literature, we focused on the conventional CI and FG amplification coefficients to maintain a long-term average power constraint at the relay. Although these two coefficients are widely used in the literature, they are simply power normalization factors and are thus not optimal in any sense. For the single-input single-output system, source adaptation schemes to maximize the achievable rate under an average power constraint were studied by Goldsmith and Varaiya in their seminal paper [4]. In [4], the source is assumed to have full CSI of the source-destination link and adapts its output power over time according to the channel conditions. Several adaptation techniques were proposed in [4] such as channel inversion and the optimal water-filling policy. Similar to the idea in [4], relay adaptation strategies have also been proposed in [2, 3] assuming that the relay can acquire knowledge of its incoming and outgoing links. Specifically, the output power at the relay in [2, 3] varies according to the CSI to maximize a given objective function while maintaining a long-term average power constraint. However, the works in [2, 3] were limited to the dual-hop scenario that ignores the source-destination link. For NAF relaying, an optimal amplification coefficient was recently obtained in [6] using the uncoded PEP criterion. This amplification coefficient is, however, just numerically optimized for the QPSK constellation and only the CDI knowledge of the channel is exploited in [6]. Although power allocation strategies among the nodes in AF networks have been previously investigated as highlighted in the previous two chapters, relay adaptation strategies for cooperative or TW AF schemes have not been addressed in the literature.

In this chapter, we study relay adaptation policies for half-duplex cooperative and TW AF systems in which the relay can exploit the CSI by means of a suitable amplification coefficient to enhance the performance. Specifically, assuming full CSI at the relay and Gaussian codebooks at the source nodes, the optimal relay functions are derived under a long-term average power constraint. We consider

L. Jiménez Rodríguez et al., *Amplify-and-Forward Relaying in Wireless Communications*, SpringerBriefs in Computer Science,
DOI 10.1007/978-3-319-17981-0_5

the achievable rate and sum rate as the performance criteria for OW and TW relaying, respectively. At first, focusing on the OAF protocol, the maximization of the achievable rate is shown to be a concave optimization problem. The KKT conditions are then used to obtain the optimal relay function in closed form. For the NAF scheme, it turns out that the maximization of the rate results in a non-concave problem. Therefore, we develop a method from which a global solution can be obtained in closed form. Similar to the OAF case, the maximization of the sum rate for the TW system is also shown to be a concave problem. Applying the KKT conditions, it is then shown that the relay adaptation solution is the root of a quartic polynomial. Numerical results indicate that the derived schemes provide significant rate improvements over the FG and CI techniques, especially in low power regions.

5.1 Problem Formulation

In this chapter, we consider again the fast fading scenario in which the transmitted codeword spans several channel realizations. In particular, we assume that the codeword is sent over N consecutive transmission frames as described by the protocols in Sect. 2.2. The OW OAF, OW NAF, and two-phase TW protocols with input-output relations (2.8), (2.1) and (2.12), respectively, are considered. The channel gains at frame i ($1 \leq i \leq N$), $\boldsymbol{h}_i = [h_0[i], h_1[i], h_2[i]]$, are assumed to be arbitrarily distributed and remain constant during at least one frame. For this faded scenario, perfect knowledge of the channel gains $\boldsymbol{h}_i$ is considered at all receivers. More importantly, the relay also has full knowledge of $\boldsymbol{h}_i$. This knowledge can be acquired via feedback from the destination in OW systems, or can be readily estimated at the relay in TW schemes.

Recall from Sect. 2.2 that two types of amplification coefficients are usually considered in literature, i.e., the FG and the CI coefficients with $\Xi = \phi$ and $\Xi = \alpha$ in (2.4) and (2.13). Given that the relay is assumed to posses full CSI in this chapter, it may use this knowledge to adapt the amplification coefficient according to the channel variations. In particular, let $z_2[i] P_r$ denote the *instantaneous* power allocated to the relay in frame i. The relay can then vary $z_2[i]$ according to $\boldsymbol{h}_i$ while maintaining a long-term average power constraint $\mathbb{E}[z_2[i]] \leq z_2$, i.e., the average power constraint is $z_2 P_r$. In this case, the relay adaptation (RA) amplification coefficient for OW and TW systems can be respectively written as

$$b[i] = \sqrt{\frac{z_2[i]}{q_1 P_s \alpha_1[i] + N_r}}, \tag{5.1}$$

and

$$b[i] = \sqrt{\frac{z_2[i]}{q_1 P_{s1} \alpha_1[i] + q_2 P_{s2} \alpha_2[i] + N_r}}, \tag{5.2}$$

where $\alpha_l[i] = |h_l[i]|^2$.

In general, the relay might adapt the amplification coefficients in (5.1) and (5.2) in order to optimize a given objective function. Among different criteria, here we consider the achievable rate or sum rate for OW and TW systems, respectively, as the performance metric similar to Chaps. 3 and 4. Certainly, one can use other design criteria to develop adaptation techniques following a similar approach [2, 3, 5]. As can be seen from (5.1) and (5.2), the problem of finding the optimal relay adaptation scheme in terms of $\boldsymbol{h}_i$, i.e., the optimal relay function $b[i]$, is equivalent to finding the optimal instantaneous power allocated to the relay at a given frame $z_2[i]$. Assuming Gaussian codebooks and that the transmitted codewords span over N frames, the optimal $z_2[i]$ can then be obtained by maximizing the achievable rate or sum rate while maintaining the average power constraint. This can be modeled as a parallel channel and the optimization problem for OW systems can be written as

$$\max_{z_2[i]\geq 0} \quad \frac{1}{N}\sum_{i=1}^{N} I|\boldsymbol{h}_i \quad \text{s.t.} \quad \frac{1}{N}\sum_{i=1}^{N} z_2[i] \leq z_2, \tag{5.3}$$

where the conditional mutual information $I|\boldsymbol{h}_i$ is given by either (2.5) or (2.9) depending on the OW protocol under consideration. Similarly, the optimization problem for TW relaying is given by

$$\max_{z_2[i]\geq 0} \frac{1}{N}\sum_{i=1}^{N} [I_1|\boldsymbol{h}_i + I_2|\boldsymbol{h}_i], \quad \text{s.t.} \quad \frac{1}{N}\sum_{i=1}^{N} z_2[i] \leq z_2, \tag{5.4}$$

with $I_1|\boldsymbol{h}_i$ as in (2.14). The systems using the conventional FG and CI coefficients can be used as benchmarks to asses the adaptation schemes in (5.3) and (5.4).

Before closing this section, it should be noted that the framework described here is general, in the sense that the average power allocation scheme among the nodes is left as the parameters $\boldsymbol{q} = [q_1, q_2, z_2]$. In the following, the main objective is to derive an optimal power adaptation scheme by means of the amplification coefficient, which can be applied to any such parameters.

5.2 Optimal Power Adaptation Schemes

In this section, we shall develop optimal power adaptation schemes by solving the problems in (5.3) and (5.4). Let $z_2[i] = z_{2,i}$ and denote the optimum value of $z_{2,i}$ as $z_{2,i}^{\star}$. We first address the adaptation problem for the OAF system, followed by that of the NAF and TW schemes.

5.2.1 OAF System

Consider first the OAF system in which the source has a power allocation of $q_1 > 0$, whereas an average power constraint of $z_2 > 0$ is imposed on the relay. Recall from Sect. 2.2 that for the OAF scheme, the source is silent in the cooperative phase with $q_2 = 0$. Substituting the RA coefficient in (5.1), the conditional achievable rate in (2.9) can be simplified to

$$I_{\text{OAF}}|\boldsymbol{h}_i = \frac{1}{2}\log\left(1 + q_1\gamma_{0,i} + \frac{q_1 z_{2,i}\gamma_{1,i}\gamma_{2,i}}{q_1\gamma_{1,i} + z_{2,i}\gamma_{2,i} + 1}\right) = \frac{1}{2}\log\left[1 + f_{\text{OAF}}(z_{2,i}, \boldsymbol{h}_i)\right], \tag{5.5}$$

where $\gamma_{0,i} = (P_s\alpha_0[i])/N_d$, $\gamma_{1,i} = (P_s\alpha_1[i])/N_r$ and $\gamma_{2,i} = (P_r\alpha_2[i])/N_d$. The optimal instantaneous power allocated to the relay for the OAF system can then be obtained by solving (5.3) with $I|\boldsymbol{h}_i$ as in (5.5). With a slight abuse of notation, let $I|\boldsymbol{h}_i = I(z_{2,i}|\boldsymbol{h}_i)$. The first and second derivatives of (5.5) are given respectively as

$$\frac{\partial I_{\text{OAF}}(z_{2,i}|\boldsymbol{h}_i)}{\partial z_{2,i}} = \frac{a_i}{2\ln(2)[c_i z_{2,i}^2 + d_i z_{2,i} + e_i]}, \tag{5.6}$$

$$\frac{\partial^2 I_{\text{OAF}}(z_{2,i}|\boldsymbol{h}_i)}{\partial z_{2,i}^2} = \frac{-a_i[2c_i z_{2,i} + d_i]}{2\ln(2)[c_i z_{2,i}^2 + d_i z_{2,i} + e_i]^2}, \tag{5.7}$$

where

$$\begin{aligned} a_i &= q_1\gamma_{1,i}\gamma_{2,i}\left(q_1\gamma_{1,i} + 1\right), \quad c_i = \gamma_{2,i}^2\left(q_1\gamma_{0,i} + q_1\gamma_{1,i} + 1\right), \\ d_i &= \gamma_{2,i}\left(q_1\gamma_{1,i} + 1\right)\left(2q_1\gamma_{0,i} + q_1\gamma_{1,i} + 2\right), \quad e_i = \left(q_1\gamma_{0,i} + 1\right)\left(q_1\gamma_{1,i} + 1\right)^2. \end{aligned} \tag{5.8}$$

Note that the second derivative in (5.7) is non-positive and independent of $z_{2,k}$, $\forall k \neq i$. The Hessian of (5.5) is then diagonal with the non-positive elements in (5.7). Given that the constraints in (5.3) are linear, this is a concave optimization problem. The optimal value of $z_{2,i}$ can then be obtained using the KKT conditions by simply finding a stationary point in the Lagrangian [7]. In particular, the Lagrangian of (5.3) can be written as

$$\mathscr{L}(\mathbf{z}_2, \lambda_1) = -\frac{1}{N}\sum_{i=1}^{N} I_{\text{OAF}}(z_{2,i}|\boldsymbol{h}_i) - \lambda_1\left(z_2 - \frac{1}{N}\sum_{i=1}^{N} z_{2,i}\right), \tag{5.9}$$

where $\mathbf{z}_2 = \{z_{2,i} \mid 1 \leq i \leq N\}$ and $\lambda_1 \geq 0$ is the Lagrange multiplier. The i-th component of the gradient of (5.9), $\nabla\mathscr{L}(\mathbf{z}_2, \lambda_1)$, is given by

$$\frac{\partial \mathscr{L}(z_2, \lambda_1)}{\partial z_{2,i}} = -\frac{1}{N}\frac{\partial I_{\mathrm{OAF}}(z_{2,i}|\boldsymbol{h}_i)}{\partial z_{2,i}} + \frac{\lambda_1}{N},$$

with the derivative as in (5.6). By equating the gradient to zero, i.e., $\nabla \mathscr{L}(z_2, \lambda_1) = \mathbf{0}$, and solving for $z_{2,i}$, the optimal instantaneous allocation for the OAF scheme can be expressed as

$$z_{2,i}^{\mathrm{OAF}} = \begin{cases} \left[\frac{-d_i + \sqrt{d_i^2 - 4c_i\left(e_i - a_i\mu_1^\star\right)}}{2c_i}\right]^+, & \gamma_{1,i}, \gamma_{2,i} > 0 \\ 0, & \text{o.w.,} \end{cases} \tag{5.10}$$

where $[x]^+ = \max\{0, x\}$, $\mu_1 = 1/[2\ln(2)\lambda_1]$ and $d_i^2 - 4c_i(e_i - a_i\mu_1) > 0$, so that the root in (5.10) is always real.

In (5.10), $\lambda_1^\star > 0$ is a unique constant that must satisfy

$$g(\lambda_1^\star) = \frac{1}{N}\sum_{i=1}^{N} z_{2,i}^\star = z_2, \tag{5.11}$$

and can be easily found using bisection. This is because the root in (5.10) is strictly decreasing with $\lambda_1 > 0$, is negative when $\lambda_1 \to +\infty$, and approaches $+\infty$ as λ_1 approaches zero from above $\lambda_1 \to 0^+$. As such, $\lim_{\lambda_1 \to 0^+} g(\lambda_1) = +\infty$, $\lim_{\lambda_1 \to +\infty} g(\lambda_1) = 0$ and $g(\lambda_1)$ is strictly decreasing with $\lambda_1 > 0$. Note also that the sum power constraint in (5.11) is tight as the objective function in (5.5) is strictly increasing with $z_{2,i}$ as long as $\exists i \in \{1, \ldots, N\}$ with $\gamma_{1,i}, \gamma_{2,i} > 0$, i.e., the first derivative in (5.6) is strictly positive.

Observe from (5.3) that the solution in (5.10) can be applied over any N parallel channels as long as the relay has knowledge of $\boldsymbol{h}_i$ for $1 \le i \le N$. However, when N large enough to capture the ergodicity of the channel $\boldsymbol{h}_i$,

$$\frac{1}{N}\sum_{i=1}^{N} I|\boldsymbol{h}_i \to \mathbb{E}[I|\boldsymbol{h}_i] = I, \quad \text{and} \quad \frac{1}{N}\sum_{i=1}^{N} z_{2,i}^\star \to \mathbb{E}[z_{2,i}^\star] = z_2.$$

In this case, $(z_{2,i}^\star, \lambda_1^\star)$ become a causal function of $\boldsymbol{h}_i$ and its distribution, rather than a non-causal function of future values of $\boldsymbol{h}_i$. It can be seen from (5.10) that $\lambda_1^\star$ also depends on the noise variances, and the power allocations at the source and relay.

5.2.2 *NAF System*

We now turn our attention to NAF system in which the source is allowed to transmit in both broadcasting and cooperative phases. As such, it can be assumed that q_1, q_2 and z_2 are strictly positive. Recall from Sect. 2.2 that the covariance matrix of the input vector to the NAF channel is given by $\boldsymbol{Q} = \mathbb{E}[\boldsymbol{x}\boldsymbol{x}^\dagger]$ and has diagonal elements $q_1 > 0$ and $q_2 > 0$. As shown in [1], when the source uses a Gaussian codebook and has no channel knowledge, correlation in the input vector reduces the average mutual information of the NAF system. Hence, similar to Chap. 4, we consider a diagonal input covariance matrix $\boldsymbol{Q}$. Substituting the RA coefficient in (5.1) into (2.5) with $q_{12} = 0$, the conditional mutual information can be written as

$$I_{\text{NAF}}|\boldsymbol{h}_i = \frac{1}{2}\log[1 + f_{\text{NAF}}(z_{2,i}, \boldsymbol{h}_i)], \tag{5.12}$$

where $f_{\text{NAF}}(z_{2,i}, \boldsymbol{h}_i) = f_{1,\text{NAF}}(z_{2,i}, \boldsymbol{h}_i)/f_{2,\text{NAF}}(z_{2,i}, \boldsymbol{h}_i)$ with

$$\begin{aligned} f_{1,\text{NAF}}(z_{2,i}, \boldsymbol{h}_i) =& q_1^2 q_2 \gamma_{0,i}^2 \gamma_{1,i} + \gamma_{0,i}(q_1 + q_2) \\ &+ q_1 \left(q_2\gamma_{0,i}^2 + q_1\gamma_{0,i}\gamma_{1,i} + z_{2,i}\gamma_{0,i}\gamma_{2,i} + z_{2,i}\gamma_{1,i}\gamma_{2,i} + q_2\gamma_{0,i}\gamma_{1,i}\right), \\ f_{2,\text{NAF}}(z_{2,i}, \boldsymbol{h}_i) =& q_1\gamma_{1,i} + z_{2,i}\gamma_{2,i} + 1. \end{aligned}$$

The optimal instantaneous power allocation at the relay can be obtained as in the previous subsection by solving (5.3) with $I|\boldsymbol{h}_i$ in (5.12). It can be easily shown that

$$\frac{\partial I_{\text{NAF}}(z_{2,i}|\boldsymbol{h}_i)}{\partial z_{2,i}} = \frac{-A_i}{2\ln(2)[C_i z_{2,i}^2 + D_i z_{2,i} + E_i]}, \tag{5.13}$$

$$\frac{\partial^2 I_{\text{NAF}}(z_{2,i}|\boldsymbol{h}_i)}{\partial z_{2,i}^2} = \frac{A_i[2C_i z_{2,i} + D_i]}{2\ln(2)[C_i z_{2,i}^2 + D_i z_{2,i} + E_i]^2}, \tag{5.14}$$

where

$$\begin{aligned} A_i &= \gamma_{2,i}\left(q_1\gamma_{1,i}+1\right)\left(q_1q_2\gamma_{0,i}^2+q_2\gamma_{0,i}-q_1\gamma_{1,i}\right), \quad C_i = \gamma_{2,i}^2\left(q_1\gamma_{0,i}+q_1\gamma_{1,i}+1\right), \\ D_i &= \gamma_{2,i}\left(q_1\gamma_{1,i} + 1\right)\left(\gamma_{0,i}[2q_1 + q_2 + q_1q_2\gamma_{0,i}] + q_1\gamma_{1,i} + 2\right), \\ E_i &= \left(q_1\gamma_{0,i} + 1\right)\left(q_2\gamma_{0,i} + 1\right)\left(q_1\gamma_{1,i} + 1\right)^2. \end{aligned} \tag{5.15}$$

Note that the parameters $E_i > 0$ and $C_i, D_i \geq 0$. However, A_i might be positive, negative or zero depending on the noise variances, power allocations at S, and channel gains. Hence, different from the OAF case, the optimization problem in (5.3) for the NAF system is not concave. In the following, we first develop an achievable upper bound on the objective function (5.12) so that a concave-optimization problem can be established. The global maximizer of (5.3) is finally obtained in closed-form.

First, observe from (5.13) that $I_{\text{NAF}}(z_{2,i}|\boldsymbol{h}_i)$ can be either strictly increasing with $z_{2,i}$ when $A_i < 0$, strictly decreasing when $A_i > 0$, or constant when $A_i = 0$. Define the sets $\text{AP} = \{i \mid A_i \geq 0\}$ and $\text{AN} = \{i \mid A_i < 0\}$. Furthermore, let $\mathscr{R} = \{\boldsymbol{z}_2 \in \mathbb{R}^N \mid z_{2,i} \geq 0, \frac{1}{N}\sum_{i=1}^{N} z_{2,i} \leq z_2\}$ be the feasible region. It is then straightforward to show that for any $z_2 \in \mathscr{R}$, the objective function in (5.3) can be upper bounded as

$$\sum_{i=1}^{N} I_{\mathrm{NAF}}(z_{2,i}|\boldsymbol{h}_i) \leq \sum_{i \in \mathrm{AN}} I_{\mathrm{NAF}}(z_{2,i}|\boldsymbol{h}_i) \leq \sum_{i \in \mathrm{AN}} I_{\mathrm{NAF}}(z_{2,i}^{\star}|\boldsymbol{h}_i). \tag{5.16}$$

In (5.16), the first upper bound is due to the fact that the objective function decreases in value with $z_{2,i} > 0$ for $i \in \mathrm{AP}$ and is achieved with equality when $z_{2,i} = 0$ $\forall i \in \mathrm{AP}$. The second inequality is the best feasible upper bound with $z_{2,i} = 0$ $\forall i \in \mathrm{AP}$ and the values of $z_{2,i}^{\star}$ for $i \in \mathrm{AN}$ are the solution to:

$$\max_{\substack{z_{2,i} \geq 0,\\ i \in \mathrm{AN}}} \frac{1}{N} \sum_{i \in \mathrm{AN}} I_{\mathrm{NAF}}(z_{2,i}|\boldsymbol{h}_i) \quad \text{s.t.} \quad \frac{1}{N} \sum_{i \in \mathrm{AN}} z_{2,i} \leq z_2. \tag{5.17}$$

Note that the above objective function is now concave since its Hessian is diagonal with the strictly negative elements in (5.14) for $A_i < 0$. Hence, due to the linearity of the constrains and the achievability of the upper bound in (5.16), any solution to (5.3) in the form of $z_{2,i} = 0$ $\forall i \in \mathrm{AP}$ and $z_{2,i} = z_{2,i}^{\star}$ $\forall i \in \mathrm{AN}$ is *globally optimal*. The problem in (5.17) can then be solved using the KKT conditions. The Lagrangian of (5.17) can be written as

$$\mathscr{L}(z_2, \lambda_2) = -\frac{1}{N} \sum_{i \in \mathrm{AN}} I_{\mathrm{NAF}}(z_{2,i}|\boldsymbol{h}_i) - \lambda_2 \left(z_2 - \frac{1}{N} \sum_{i \in \mathrm{AN}} z_{2,i} \right).$$

By equating the gradient of the above Lagrangian to zero and solving for $z_{2,i}$, the optimal instantaneous allocation $z_{2,i}^{\star}$ for the NAF protocol and the solution to (5.3) is finally expressed as

$$z_{2,i}^{\mathrm{NAF}} = \begin{cases} \left[\frac{-D_i + \sqrt{D_i^2 - 4C_i(E_i + A_i \mu_2^{\star})}}{2C_i} \right]^{+}, & A_i < 0 \ (i \in \mathrm{AN}) \\ 0, & A_i \geq 0 \ (i \in \mathrm{AP}), \end{cases} \tag{5.18}$$

In (5.18), $\mu_2 = 1/[2\ln(2)\lambda_2]$ and $D_i^2 - 4C_i(E_i + A_i \mu_2^{\star}) > 0$ for $A_i < 0$ so that $z_{2,i}^{\star}$ is always real. Furthermore, $\lambda_2^{\star} > 0$ is a unique constant which satisfies (5.11) and can be easily found using bisection as long as the set $\mathrm{AN} \neq \emptyset$. This is because, similar to (5.10), the root (5.18) is strictly decreasing with $\lambda_2 > 0$, is negative when $\lambda_2 \to +\infty$, and approaches $+\infty$ as $\lambda_2 \to 0^{+}$. The sum power constraint in (5.11) is again tight as the objective function in (5.17) is strictly increasing with $z_{2,i}$ for $i \in \mathrm{AN}$.

5.2.3 TW System

Consider now the TW system in which both source nodes want to exchange information. For the system using the RA coefficient in (5.2), the conditional achievable rate in the $S_k \to S_j$ direction ($j \in \{1,2\}, k = 3-j$) can be expressed from (2.14) as

$$I_{j,\mathrm{TW}}|\boldsymbol{h}_i = \frac{1}{2}\log\left(1 + \frac{q_k z_{2,i}\gamma_{1,i}\gamma_{2,i}}{q_1\gamma_{1,i} + q_2\gamma_{2,i} + z_{2,i}\gamma_{j,i} + 1}\right), \tag{5.19}$$

where for notational convenience we assume that $P_{sj} = P_r$ and $N_{dj} = N_r$ so that $\gamma_{j,i} = (P_{sj}\alpha_j[i])/N_r = (P_r\alpha_j[i])/N_{dj}$. The conditional sum rate is then given by $I_{\mathrm{sum}}|\boldsymbol{h}_i = I_1|\boldsymbol{h}_i + I_2|\boldsymbol{h}_i$. The optimal power allocation that maximizes the sum rate can be obtained by solving (5.4) using $I_j|\boldsymbol{h}_i$ in (5.19). The first and second derivatives of (5.19) are given by

$$\frac{\partial I_{j,\mathrm{TW}}(z_{2,i}|\boldsymbol{h}_i)}{\partial z_{2,i}} = \frac{A'_j}{2\ln(2)P_j(z_{2,i})}, \quad \frac{\partial^2 I_{j,\mathrm{TW}}(z_{2,i}|\boldsymbol{h}_i)}{\partial^2 z_{2,i}} = \frac{-A'_j[2B'_j z_{2,i} + D'_j]}{2\ln(2)P_j^2(z_{2,i})}, \tag{5.20}$$

where the second order polynomial

$$P_j(z_{2,i}) = B'_j z_{2,i}^2 + D'_j z_{2,i} + E', \tag{5.21}$$

with

$$\begin{aligned} A'_j &= q_k\gamma_{1,i}\gamma_{2,i}(q_1\gamma_{1,i} + q_2\gamma_{2,i} + 1), \quad B'_j = \gamma_{j,i}^2(q_k\gamma_{k,i} + 1), \\ D'_j &= \gamma_{j,i}(q_k\gamma_{k,i} + 2)(q_1\gamma_{1,i} + q_2\gamma_{2,i} + 1), \quad E' = (q_1\gamma_{1,i} + q_2\gamma_{2,i} + 1)^2. \end{aligned} \tag{5.22}$$

Note that the second derivative in (5.20) is non-positive. Similar to the OAF case, the problem in (5.4) is concave for the TW system and the optimal $z_{2,i}^\star$ can be obtained using the Lagrangian method. The Lagrangian of (5.4) can be written as

$$\mathscr{L}(z_2, \lambda_3) = -\frac{1}{N}\sum_{i=1}^{N}[I_{1,\mathrm{TW}}(z_{2,i}|\boldsymbol{h}_i) + I_{2,\mathrm{TW}}(z_{2,i}|\boldsymbol{h}_i)] - \lambda_3\left(z_2 - \frac{1}{N}\sum_{i=1}^{N} z_{2,i}\right), \tag{5.23}$$

and its gradient is given by

$$\frac{\partial\mathscr{L}(z_2,\lambda_3)}{\partial z_{2,i}} = -\frac{1}{N}\left[\frac{\partial I_{1,\mathrm{TW}}(z_{2,i}|\boldsymbol{h}_i)}{\partial z_{2,i}} + \frac{\partial I_{2,\mathrm{TW}}(z_{2,i}|\boldsymbol{h}_i)}{\partial z_{2,i}}\right] + \frac{\lambda_3}{N},$$

with the derivatives as in (5.20). By equating the above derivative to zero, it can be shown that the stationary point of the Lagrangian must satisfy the following:

$$P(z_{2,i}, \mu_3) = P_1(z_{2,i})P_2(z_{2,i}) - \mu_3 A_1' P_2(z_{2,i}) - \mu_3 A_2' P_1(z_{2,i}) = 0, \tag{5.24}$$

for $1 \leq i \leq N$, where $\mu_3 = 1/[2\ln(2)\lambda_3]$ and $P_j(\cdot)$ is given by (5.21). The above function is a quartic polynomial on $z_{2,i}$ and has at most four real roots. Since (5.4) is a concave problem, at most one of these roots is real and positive.

Rather than writing the cumbersome equation for the root of a quartic polynomial, let $r_i(\mu_3)$ be the largest real root of (5.24) at frame i for a given μ_3, i.e., $P(r_i(\mu_3), \mu_3) = 0$. When $\gamma_{1,i} = 0$ or $\gamma_{2,i} = 0$, $P(z_{2,i}, \mu_3)$ becomes a quadratic with a single real negative root of multiplicity two. From (5.24) and the concavity of (5.4), the optimal power allocation and the solution to (5.4) can then be expressed as

$$z_{2,i}^{\mathrm{TW}} = \begin{cases} [r_i(\mu_3^\star)]^+, & \gamma_{1,i}, \gamma_{2,i} > 0 \\ 0, & \text{o.w.} \end{cases} \tag{5.25}$$

It can be easily shown that $\lim_{\lambda_3 \to +\infty} r_i(\mu_3) < 0$, $\lim_{\lambda_3 \to 0^+} r_i(\mu_3) = +\infty$ and $r_i(\mu_3)$ is strictly decreasing with $\lambda_3 > 0$. Thus, as in the NAF and OAF cases, $\lambda_3^\star > 0$ in (5.25) can be found using bisection. The sum power constraint in (5.4) is again tight as the first derivative in (5.20) is strictly positive for $\gamma_{1,i}, \gamma_{2,i} > 0$.

5.3 Illustrative Examples

In this section, simulation results are presented to quantify the gain of the proposed schemes. The two conventional systems using the FG and CI coefficients are used as a baseline for comparison. The channel gains are assumed to be independent with Rayleigh-distributed magnitude as $h_l[i] \sim \mathscr{CN}(0, \phi_l)$. Similar to Chaps. 3 and 4, we adopt the linear network model so that $\phi_0 = 1$, $\phi_1 = 1/d^\nu$ and $\phi_2 = 1/(1-d)^\nu$, where $0 \leq d \leq 1$ is the normalized $S_1 - R$ distance and ν is the pathloss exponent. For all simulations, $\nu = 3$, $d = 0.5$ and unit noise power ($N_0 = 1$) at the three nodes is considered. In addition, equal power allocation is assumed with $P_s = P_r = P_{sj} = P_t$, $q_1 = q_2 = z_2 = 2/3$ for TW/NAF and $q_1 = z_2 = 1$ for OAF.

Figure 5.1 shows the achievable rates of the OAF system in (5.5) against P_t/N_0 for three different techniques. These include the FG coefficient in (2.4) with $\Xi = \phi$, the CI coefficient in (2.4) with $\Xi = \alpha$, and the RA coefficient in (5.1) with $z_2[i]$ in (5.10). In this and all subsequent figures, the rates are averaged over 10^6 channel realizations and are normalized by the rate of the FG technique. Observe from Fig. 5.1 that the proposed RA technique provides the best performance for all values of P_t/N_0. The gains over the CI and FG schemes are significant in low P_t/N_0 regions. For instance, a 1.4× increase in rate can be observed at $P_t/N_0 = -20$ dB.

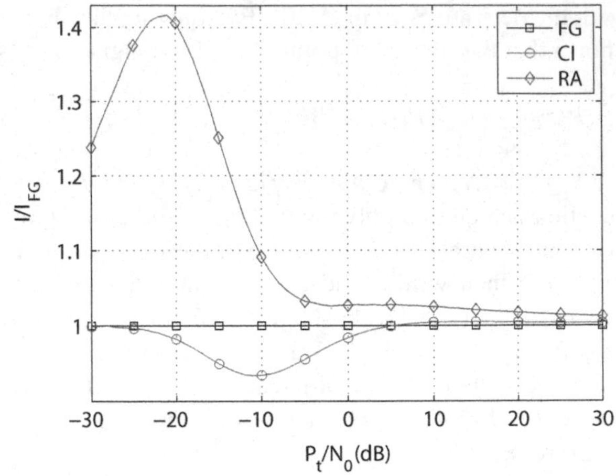

Fig. 5.1 Normalized rates for the OAF system using different power adaptation schemes

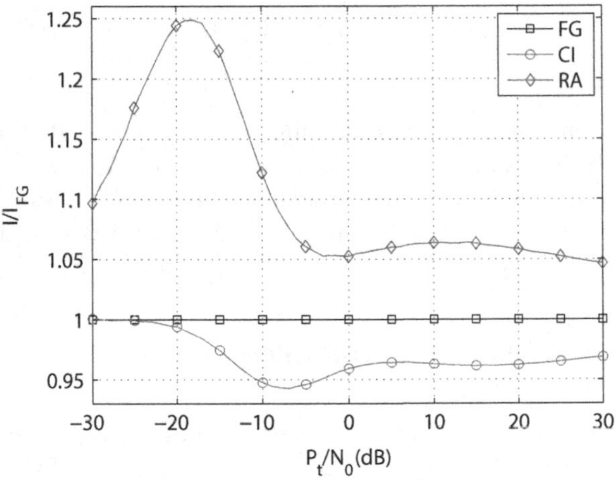

Fig. 5.2 Normalized rates for the NAF system using different power adaptation schemes

The normalized achievable rates of the NAF system in (5.12) are shown in Fig. 5.2 for different schemes. As expected, the proposed RA technique in (5.1) and (5.18) outperforms the CI and FG coefficients for the entire range of P_t/N_0 in Fig. 5.2. Similar to the OAF case, the gains are more significant in low P_t regions. For example, a 1.25× gain is achieved in Fig. 5.2 at $P_t/N_0 = -20$ dB.

Figure 5.3 shows the normalized sum rates of the TW system using the FG/CI coefficients in (2.13), and the RA coefficient in (5.2) and (5.25). Similar to the previous cases, it can be seen from Fig. 5.3 that the proposed RA technique provides impressive gains over the FG and CI techniques, specially in low power regions.

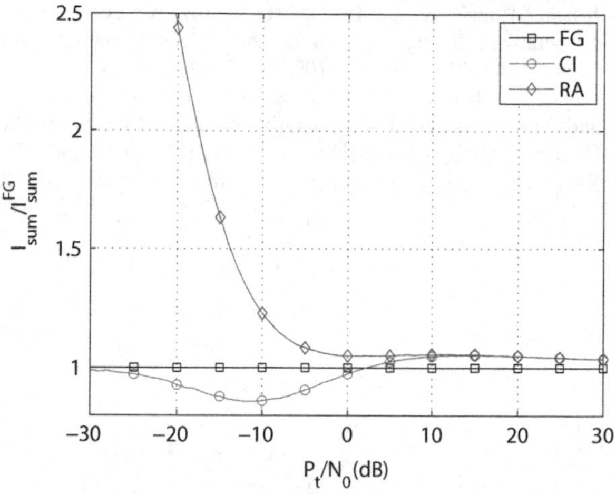

Fig. 5.3 Normalized sum rates of the TW system using different power adaptation schemes

Specifically, a two-fold increase over the FG system can be seen at −17 dB. However, for $P_t/N_0 > 10$ dB, the sum rates of the CI and RA systems converge and do not significantly outperform that of the FG system.

5.4 Concluding Remarks

This chapter considered the design of optimal relay adaptation schemes over OAF, NAF and TWAF channels. By assuming Gaussian codebooks at the source nodes and full CSI at the relay, optimal power amplifications were derived to maximize the achievable rate or sum rate of the considered systems. Simulation results were finally provided to demonstrate the advantage of the proposed relay adaptation system over conventional AF systems using either the FG or the CI coefficient.

References

1. Ding, Y., Zhang, J.K., Wong, K.M.: Ergodic channel capacities for the amplify-and-forward half-duplex cooperative systems. IEEE Trans. Inf. Theory **55**(2), 713–730 (2009). Doi:10.1109/TIT.2008.2009822
2. Gatzianas, M., Georgiadis, L., Karagiannidis, G.: Optimal relay control in power-constrained dual-hop transmissions over arbitrary fading channels. In: Proceedings of IEEE International Conference Communications, vol. 10, pp. 4543–4548 (2006). Doi:10.1109/ICC.2006.255355
3. Gatzianas, M., Georgiadis, L., Karagiannidis, G.: Gain adaptation policies for dual-hop nonregenerative relayed systems. IEEE Trans. Commun. **55**(8), 1472–1477 (2007). Doi:10.1109/TCOMM.2007.902530
4. Goldsmith, A., Varaiya, P.: Capacity of fading channels with channel side information. IEEE Trans. Inf. Theory **43**(6), 1986–1992 (1997). Doi:10.1109/18.641562

5. Jiménez-Rodríguez, L., Tran, N.H., Helmy, A., Le-Ngoc, T.: Optimal power adaptation for cooperative AF relaying with channel side information. IEEE Trans. Veh. Technol. **62**(7), 3164–3174 (2013). Doi:10.1109/TVT.2013.2250321
6. Liang, J., Zhang, J.K., Wong, K.: Optimum 4-QAM relay amplification for the amplify-and-forward half-duplex cooperative wireless system. In: Proceedings of IEEE International Conference Signal Processing, pp. 2423–2426 (2010). Doi:10.1109/ICOSP.2010.5655086
7. Wright, S., Nocedal, J.: Numerical Optimization. Springer, NewYork (2006)

Chapter 6
Half-Duplex AF Relaying: Error Performance and Precoding

In previous chapters, we addressed the capacity of several AF relay systems assuming Gaussian codebooks at the source nodes. Besides the rate performance with Gaussian inputs, it is also important to design practical coding schemes using finite constellations. In this regard, several techniques have been proposed in the literature to exploit the diversity of AF systems over fast fading channels. For instance, by applying the idea of signal space diversity [3] to uncoded NAF systems via a precoding matrix, it was shown in [5, 6] that full cooperative diversity can be achieved. By further incorporating such a precoding technique into a bit-interleaved coded modulation system with iterative decoding (BICM-ID), references [10, 13] showed that both time and cooperative diversities can be exploited simultaneously. In particular, it was demonstrated in [10, 13] that the diversity gain function of the considered NAF-BICM-ID system is d_H-th power of that of the uncoded full-diversity cooperative system in [5, 6]. Since the precoders in [10, 13] are restricted to a single cooperative frame, the maximum achievable diversity order is constrained by the minimum Hamming distance d_H of the outer code. Without this restriction, higher degrees of freedom, and consequently, larger time diversity order, can be attained and incorporated into the cooperative system for performance improvement.

Usually, diversity benefits offered by relaying can only be observed in the error-floor region when the system operates at a sufficiently high SNR. For an ergodic fading relay channel, designing a practical coding strategy that can approach closely to its fundamental limit in the turbo pinch-off or waterfall region is also particularly important. To the best of our knowledge, most of the designs available so far are made to DF relaying [9, 17, 18]. Despite its practicality, there is not much advancement in designing a good coding scheme for NAF systems.

L. Jiménez Rodríguez et al., *Amplify-and-Forward Relaying in Wireless Communications*, SpringerBriefs in Computer Science,
DOI 10.1007/978-3-319-17981-0_6

Motivated by the above discussions, this chapter proposes a precoding scheme over multiple cooperative frames for a BICM system over a half-duplex NAF relay channel. We focus on both diversity improvement in the error-floor region and near-capacity performance in the turbo pinch-off region. Specifically:

- Focusing on the error-floor region, we show that the multiple-frame precoded system can provide a diversity gain function that is N-th power of that of the previously considered NAF-BICM-ID system in [10], where N is the number of precoded cooperative frames. An optimal class of precoders is then derived based on the asymptotic coding gain. This class indicates that the source should transmit a superposition of all symbols in the broadcasting phases, while being silent in all cooperative phases to optimize the asymptotic performance. A design criterion is also developed to find optimal superposition angles for good convergence behavior. A pragmatic approach is then proposed to obtain good rotation angles. Analytical and simulation results indicate that the proposed precoders achieve higher diversity orders and provide significant coding gains over the single-frame precoding scheme.
- In the turbo pinch-off region, we demonstrate that a concatenation of multi-D mapping [14], multiple-frame precoding and a simple outer binary code can be used to achieve near-capacity performance. We first show that for a given unitary and full diversity precoder, the optimal multi-D labeling for NAF relaying shall maximize the average Euclidean distance between all pairs in the multi-D constellation whose labels differ in only 1 bit. The extrinsic information transfer (EXIT) charts are then used to match the outer code, the multi-D mapping, and the precoder. For various spectral efficiencies, we obtain a BER of 10^{-5} or lower at a SNR that is 0.45 dB–1 dB below the achievable rate with finite constellations and within 1.55 dB–1.95 dB from the ergodic capacity with Gaussian inputs.

6.1 System Model

The general block diagram of the proposed NAF-BICM-ID system using a $2N \times 2N$ precoder $\boldsymbol{G}$ over N cooperative frames is shown in Fig. 6.1. The information sequence $\underline{\boldsymbol{u}}$ is first encoded into a coded sequence $\underline{\boldsymbol{c}}$. This coded sequence is then interleaved by a bit-interleaver Π to become the interleaved sequence $\underline{\boldsymbol{v}}$. Each group of $2Nm_c$ bits in $\underline{\boldsymbol{v}}$ is mapped to a signal $\boldsymbol{s} = [s_1, s_2, \ldots, s_{2N}]^\top$ in the complex $2N$-dimensional ($2N$-D) constellation Ψ according to a mapping rule ξ. Each component s_k is in a 1-D unit-energy constellation Ω, such as QPSK or QAM, of size 2^{m_c}. In general, the mapping rule ξ can be implemented independently in Ω for each component s_k or might span the entire N cooperative frames, which is referred to as multi-D mapping [14]. The symbol $\boldsymbol{s} \in \Psi$ is then rotated by a $2N \times 2N$ complex precoder $\boldsymbol{G}$ with entries $\{g_{i,k}\}$ ($1 \leq i,k \leq 2N$). Each rotated symbol $\boldsymbol{x} = \boldsymbol{G}\boldsymbol{s}$ is then transmitted via N frames over the NAF channel as described in (2.1). Let $\boldsymbol{x} = \boldsymbol{G}\boldsymbol{s} = [\boldsymbol{x}_1^\top, \ldots, \boldsymbol{x}_N^\top]^\top$. Then, the 2×1 vector

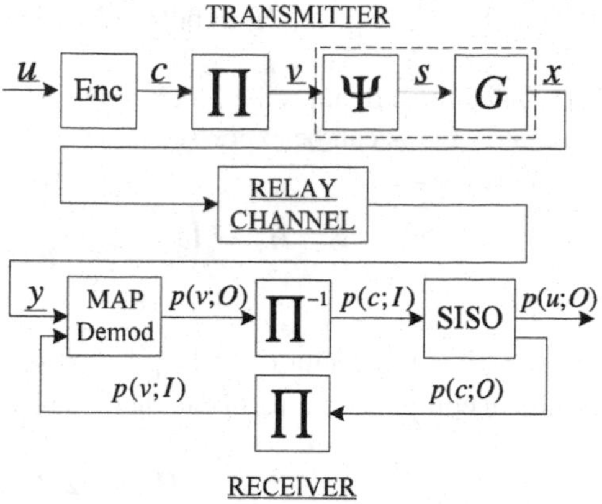

Fig. 6.1 Block diagram of a NAF-BICM-ID system using a precoder $\boldsymbol{G}$ over multiple cooperative frames

$\boldsymbol{x}_i = [x_{1,i}, x_{2,i}]^\top = \boldsymbol{G}_i\boldsymbol{s}$ is transmitted in the i-th cooperative frame and the sub-precoder $\boldsymbol{G}_i$ is of size $2 \times 2N$. At the destination, the received vector $\boldsymbol{y}$ is demodulated and decoded in an iterative manner. The receiver consists of a soft-output demodulator that follows the maximum *a posteriori* probability (MAP) algorithm similar to [15], and a soft-input soft-output (SISO) channel decoder which uses the MAP algorithm in [2].

Let the channel gains at frame i be given by $\boldsymbol{h}_i = [h_{0,i}, h_{1,i}, h_{2,i}]$ according to Fig. 2.1. In this chapter, we again consider the block fast fading scenario in which the transmitted codeword spans several realizations. In particular, the channel gains remain constant during one cooperative frame but change independently from one frame to another. Furthermore, we assume that the destination has perfect knowledge of these channel gains, while the relay only knows the second order statistics of the S-R channel. To concentrate on the code design, unit-variance Rayleigh fading $\mathscr{CN}(0, 1)$ and equal noise variances at relay and destination $N_d = N_r = N_0$ are considered. Denote the covariance matrix of the transmitted vector as $\boldsymbol{Q} = \mathbb{E}[\boldsymbol{x}\boldsymbol{x}^\dagger] = \boldsymbol{G}\boldsymbol{G}^\dagger$. From (2.2), the power transmitted by the source is then $P_s \cdot \text{tr}(\boldsymbol{Q}) = P_s \cdot \text{tr}(\boldsymbol{G}\boldsymbol{G}^\dagger)$ per cooperative frame or $P_s \cdot \text{tr}(\boldsymbol{Q})/2N$ per symbol period. Here, we assume that $\text{tr}(\boldsymbol{Q}) = \text{tr}(\boldsymbol{G}\boldsymbol{G}^\dagger) = \sum_{i=1}^{2N}\sum_{k=1}^{2N}|g_{i,k}|^2 = 2N$ so that the transmitted power is P_s per symbol period. Allocating a power of $P_r = P_s$ per cooperative frame to the relay, the FG amplification coefficient at frame i can be written as $b_i = \sqrt{1/[\eta_i P_s + N_0]}$, where $\eta_i = \sum_{l=1}^{2N}|g_{2i-1,l}|^2$ for $1 \le i \le N$. Hereafter, a group of N cooperative frames shall be referred to as a super-frame. Given these constraints, the source spends a total power of $2NP_s$, while the relay uses NP_s in any given super-frame.

By stacking the received signals in (2.2) for N frames and whitening the noise components, the $2N \times 1$ received vector at D can be written as

$$\boldsymbol{y} = \sqrt{P_s}\boldsymbol{H}\boldsymbol{x} + \boldsymbol{n}, \tag{6.1}$$

where $\boldsymbol{n} \sim \mathscr{CN}(\boldsymbol{0}, N_0\boldsymbol{I}_{2N})$, and the $2N \times 2N$ equivalent channel matrix is given by

$$\boldsymbol{H} = \text{bdiag}\,(\boldsymbol{H}_1, \boldsymbol{H}_2, \ldots, \boldsymbol{H}_N)\,. \tag{6.2}$$

In (6.2), bdiag(·) denotes the block-diagonal matrix function and the 2×2 channel sub-matrix $\boldsymbol{H}_i$ is can be expressed as

$$\boldsymbol{H}_i = \begin{pmatrix} h_{0,i} & 0 \\ \upsilon_i \sqrt{P_s} b_i h_{2,i} h_{1,i} & \upsilon_i h_{0,i} \end{pmatrix},$$

with the whitening factor $\upsilon_i = 1/\sqrt{1 + P_s b_i^2 |h_{2,i}|^2}$. For mathematical convenience, the signal component in (6.1) can then be rewritten as in [5]:

$$\boldsymbol{H}\boldsymbol{x} = \boldsymbol{\Sigma}\boldsymbol{X}\boldsymbol{T}\boldsymbol{h}_{01}, \tag{6.3}$$

where $\boldsymbol{\Sigma} = \text{bdiag}\,(\boldsymbol{\Sigma}_1, \boldsymbol{\Sigma}_2, \ldots, \boldsymbol{\Sigma}_N)$ with $\boldsymbol{\Sigma}_i = \begin{pmatrix} 1 & 0 \\ 1 & \upsilon_i \end{pmatrix}$, $\boldsymbol{X} = \text{bdiag}\,(\boldsymbol{X}_1, \boldsymbol{X}_2, \ldots, \boldsymbol{X}_N)$ with $\boldsymbol{X}_i = \begin{pmatrix} x_{1,i} & 0 \\ x_{2,i} & x_{1,i} \end{pmatrix}$, $\boldsymbol{T} = \text{bdiag}\,(\boldsymbol{T}_1, \boldsymbol{T}_2, \ldots, \boldsymbol{T}_N)$ with $\boldsymbol{T}_i = \text{bdiag}\left(1, \sqrt{P_s} b_i h_{2,i}\right)$, and $\boldsymbol{h}_{01}^{\top} = \left(\boldsymbol{h}_{01,1}^{\top}, \ldots, \boldsymbol{h}_{01,N}^{\top}\right)$ with $\boldsymbol{h}_{01,i} = (h_{0,i}, h_{1,i})^{\top}$.

6.2 Performance Analysis

This section presents a union bound to the BER for the NAF-BICM-ID system precoding over multiple cooperative frames. This bound will be used in the next sections for the designs in both error-floor and turbo pinch-off areas. In general, the union bound to the BER P_{b} for the NAF-BICM-ID system using a rate-k_c/n_c outer code is given by

$$P_{\text{b}} \leq \frac{1}{k_{\text{c}}} \sum_{d_{\text{H}}}^{\infty} c_d \, f(d, \Psi, \xi, \boldsymbol{G}), \tag{6.4}$$

where c_d is the total information weight of all error events at Hamming distance d and d_{H} is the free Hamming distance of the code. The function $f(d, \Psi, \xi, \boldsymbol{G})$ is the average PEP between two codewords and can be computed as follows.

First, let e denote the super-frame index. Then, let $\underline{c}$ and $\underline{\check{c}}$ denote respectively the input and estimated sequences with Hamming distance d between them. These binary sequences correspond to the sequences $\underline{s}$ and $\underline{\check{s}}$, whose elements are $2N$-D symbols in Ψ. Without loss of generality, assume $\underline{c}$ and $\underline{\check{c}}$ differ in the first d bits. Hence, $\underline{s}$ and $\underline{\check{s}}$ can be redefined as sequences of d complex $2N$-D symbols as $\underline{s} = [\boldsymbol{s}_{(1)}, \ldots, \boldsymbol{s}_{(d)}]$ and $\underline{\check{s}} = [\check{\boldsymbol{s}}_{(1)}, \ldots, \check{\boldsymbol{s}}_{(d)}]$. Also, let $\underline{\boldsymbol{H}} = [\boldsymbol{H}_{(1)}, \ldots, \boldsymbol{H}_{(d)}]$, where $\boldsymbol{H}_{(e)}$ represents the channel matrix in (6.2) at super-frame e affecting the transmitted symbol $\boldsymbol{s}_{(e)}$, $1 \leq e \leq d$. Then, the PEP conditioned on $\underline{\boldsymbol{H}}$ can be computed from (6.1) as

$$\mathbb{P}(\underline{s} \to \underline{\check{s}}|\underline{H}) = Q\left(\sqrt{\frac{1}{2N_0}\sum_{e=1}^{d} d^2(\boldsymbol{s}_{(e)}, \check{\boldsymbol{s}}_{(e)}|\boldsymbol{H}_{(e)})}\right), \tag{6.5}$$

where $Q(\cdot)$ is the Gaussian probability integral and $d^2(\boldsymbol{s}_{(e)}, \check{\boldsymbol{s}}_{(e)}|\boldsymbol{H}_{(e)})$ is the squared Euclidean distance between $\boldsymbol{s}_{(e)}$ and $\check{\boldsymbol{s}}_{(e)}$ conditioned on $\boldsymbol{H}_{(e)}$ and in the absence of noise:

$$d^2(\boldsymbol{s}_{(e)}, \check{\boldsymbol{s}}_{(e)}|\boldsymbol{H}_{(e)}) = P_s\|\boldsymbol{H}_{(e)}\boldsymbol{G}(\boldsymbol{s}_{(e)} - \check{\boldsymbol{s}}_{(e)})\|^2. \tag{6.6}$$

Using the alternate representation of $Q(x) = \frac{1}{\pi}\int_0^{\pi/2}\exp\left(-x^2/[2\sin^2\theta]\right)d\theta$ and by averaging over each $\boldsymbol{H}_{(e)}$ in $\underline{\boldsymbol{H}}$, the unconditional PEP is given by

$$\mathbb{P}(\underline{s} \to \underline{\check{s}}) = \frac{1}{\pi}\int_0^{\pi/2}\left[\prod_{e=1}^{d}\Delta_\theta\left(\boldsymbol{s}_{(e)}, \check{\boldsymbol{s}}_{(e)}\right)\right]d\theta, \tag{6.7}$$

where

$$\Delta_\theta\left(\boldsymbol{s}_{(e)}, \check{\boldsymbol{s}}_{(e)}\right) = \mathbb{E}_{\boldsymbol{H}_{(e)}}\left[\exp\left(-c_\theta\|\boldsymbol{H}_{(e)}\boldsymbol{G}(\boldsymbol{s}_{(e)} - \check{\boldsymbol{s}}_{(e)})\|^2\right)\right], \tag{6.8}$$

with $c_\theta = P_s/[4N_0\sin^2\theta]$. Then, from the alternate matrix model in (6.3), the distance in (6.6) can be re-written as

$$d^2(\boldsymbol{s}_{(e)}, \check{\boldsymbol{s}}_{(e)}|\boldsymbol{H}_{(e)}) = P_s\boldsymbol{h}_{01}^{(e)\dagger}\boldsymbol{T}_{(e)}^{\dagger}\boldsymbol{U}_{(e)}^{\dagger}\boldsymbol{\Sigma}_{(e)}^{\dagger}\boldsymbol{\Sigma}_{(e)}\boldsymbol{U}_{(e)}\boldsymbol{T}_{(e)}\boldsymbol{h}_{01}^{(e)},$$

where $\boldsymbol{U}_{(e)} = \boldsymbol{X}_{(e)} - \check{\boldsymbol{X}}_{(e)} = \text{bdiag}\left(\boldsymbol{U}_1^{(e)}, \ldots, \boldsymbol{U}_N^{(e)}\right)$ with $\boldsymbol{U}_i^{(e)} = \begin{pmatrix} u_{1,i}^{(e)} & 0 \\ u_{2,i}^{(e)} & u_{1,i}^{(e)} \end{pmatrix}$. Using the fact that for $\boldsymbol{a} \sim \mathscr{CN}(\boldsymbol{0}, \boldsymbol{\Phi})$ and a Hermitian matrix $\boldsymbol{A}$, $\mathbb{E}[\exp(-\boldsymbol{a}^H\boldsymbol{A}\boldsymbol{a})] = 1/\det(\boldsymbol{I} + \boldsymbol{\Phi}\boldsymbol{A})$ to average over $\boldsymbol{h}_{01}^{(e)}$, (6.8) can be simplified as in [5] to

$$\Delta_\theta\left(\boldsymbol{s}_{(e)}, \check{\boldsymbol{s}}_{(e)}\right) = \mathbb{E}_{h_{2,1}^{(e)},\ldots,h_{2,N}^{(e)}}\left[1/\det(\boldsymbol{A}_{(e)})\right],$$

where $\boldsymbol{A}_{(e)} = \boldsymbol{I}_{2N} + c_\theta\boldsymbol{T}_{(e)}^{\dagger}\boldsymbol{U}_{(e)}^{\dagger}\boldsymbol{\Sigma}_{(e)}^{\dagger}\boldsymbol{\Sigma}_{(e)}\boldsymbol{U}_{(e)}\boldsymbol{T}_{(e)}$. It is easy to verify that $\boldsymbol{A}_{(e)}$ is block diagonal and therefore can be written as $\boldsymbol{A}_{(e)} = \text{bdiag}\left(\boldsymbol{A}_1^{(e)}, \ldots, \boldsymbol{A}_N^{(e)}\right)$, where

$$A_i^{(e)} = I_2 + c_\theta \begin{pmatrix} |u_{1,i}^{(e)}|^2 + \upsilon_i^{(e)^2}|u_{2,i}^{(e)}|^2 & \upsilon_i^{(e)^2}\sqrt{P_s}b_i h_{2,i}^{(e)} u_{2,i}^{(e)^H} u_{1,i}^{(e)} \\ \upsilon_i^{(e)^2}\sqrt{P_s}b_i h_{2,i}^{(e)} u_{1,i}^{(e)^H} u_{2,i}^{(e)} & \upsilon_i^{(e)^2} P_s b_i^2 |h_{2,i}^{(e)}|^2 |u_{1,i}^{(e)}|^2 \end{pmatrix}.$$

Given that $A_{(e)}$ is block diagonal, its determinant is simply the product of each determinant of $A_i^{(e)}$. Then, $\Delta_\theta(s_{(e)}, \breve{s}_{(e)})$ can be simplified to

$$\Delta_\theta\left(s_{(e)}, \breve{s}_{(e)}\right) = \prod_{i=1}^{N} \Delta_{\theta,i}\left(s_{(e)}, \breve{s}_{(e)}\right), \tag{6.9}$$

where $\Delta_{\theta,i}\left(s_{(e)}, \breve{s}_{(e)}\right) = \mathbb{E}_{h_{2,i}^{(e)}}[\det(A_i^{(e)})^{-1}]$ has already been computed in [10] and can be expressed in a closed-form as

$$\Delta_{\theta,i}\left(s_{(e)}, \breve{s}_{(e)}\right) = \frac{1}{(c_\theta|u_{1,i}^{(e)}|^2+1)^2}\left\{1 + \left(\frac{1}{P_s b_i^2} - a\right)\mathscr{I}(a)\right\}. \tag{6.10}$$

In (6.10), $a = (c_\theta\|u_i^{(e)}\|^2 + 1)/[P_s b_i^2(c_\theta|u_{1,i}^{(e)}|^2 + 1)^2]$; $\mathscr{I}(x) = \exp(x)E_1(x)$ with $E_1(x)$ the exponential integral as in (4.4); and $\|u_i^{(e)}\|^2 = |u_{1,i}^{(e)}|^2 + |u_{2,i}^{(e)}|^2$ with $u_i^{(e)} = [u_{1,i}^{(e)}, u_{2,i}^{(e)}]^\top = G_i(s_{(e)} - \breve{s}_{(e)}) = G_i\epsilon_{(e)}$.

The PEP can finally be obtained by substituting (6.9) into (6.7). The function $f(d, \Psi, \xi, G)$ can then be upper bounded by averaging over all possible cases of PEPs in (6.7). Using the error-free feed-back bound [15], one has perfect a priori information of the coded bits fed back to the demodulator. As such, the other coded bits carried by the transmitted symbol can be assumed to be known perfectly and the error happens only when the labels of $s_{(e)}$ and $\breve{s}_{(e)}$ differ by only 1 bit. By further using the fact that the channel changes independently with e, the upper bound on $f(d, \Psi, \xi, G)$ can be simply obtained by averaging over the constellation Ψ as

$$f(d, \Psi, \xi, G) \le \frac{1}{\pi}\int_0^{\pi/2}\Big[\underbrace{\mathbb{E}_{s,p}\{\Delta_\theta(s, p)\}}_{\gamma_\theta(\Psi,\xi,G)}\Big]^d d\theta = \frac{1}{\pi}\int_0^{\pi/2}[\gamma_\theta(\Psi, \xi, G)]^d\, d\theta. \tag{6.11}$$

In (6.11), the expectation is taken over all 1-bit neighbors in Ψ, i.e., over all pairs $s, p \in \Psi$ whose labels differ in only 1 bit. Furthermore, $\Delta_\theta(s, p)$ is calculated as in (6.9), with s and p substituted for $s_{(e)}$ and $\breve{s}_{(e)}$. The expectation in (6.11) is then expressed as

$$\gamma_\theta(\Psi, \xi, G) = \frac{1}{2Nm_c 2^{2Nm_c}}\sum_{s\in\Psi}\sum_{k=1}^{2Nm_c}\left[\prod_{i=1}^{N}\Delta_{\theta,i}(s, p)\right], \tag{6.12}$$

where $\boldsymbol{u}_i$ in $\Delta_{\theta,i}(\boldsymbol{s},\boldsymbol{p})$ is given as $\boldsymbol{u}_i = [u_{1,i}^\top, u_{2,i}^\top]^\top = \boldsymbol{G}_i(\boldsymbol{s}-\boldsymbol{p}) = \boldsymbol{G}_i\boldsymbol{\epsilon}$. The perfect feed-back bound can then be obtained from (6.4) by substituting $f(d,\Psi,\xi,\boldsymbol{G})$ as in (6.11), $\gamma_\theta(\Psi,\xi,\boldsymbol{G})$ as in (6.12) and $\Delta_{\theta,i}(\boldsymbol{s},\boldsymbol{p})$ as in (6.10). As shall be shown in Sect. 6.5, this bound is tight at practical BER levels and is therefore useful to predict the error performance of the considered NAF-BICM-ID system.

6.3 Diversity Benefits in Error-Floor Regions

Although the error bound derived earlier is helpful to predict the error performance, it does not provide an insight about the effect of the proposed precoding scheme to the performance. In this section, by further simplifying the bound, we examine the diversity benefits offered by the multiple-frame precoding technique in the error-floor region. The optimal precoding scheme is then developed to minimize the error performance in the error-floor area.

6.3.1 Diversity and Coding Gain Functions

By applying the Chernoff bound $Q(\sqrt{2x}) < [1/2]\exp(-x)$ for $x > 0$, $f(d,\Psi,\xi,\boldsymbol{G})$ in (6.11) can further be upper bounded as

$$f(d,\Psi,\xi,\boldsymbol{G}) \le \frac{1}{\pi}\int_0^{\pi/2} [\gamma_\theta(\Psi,\xi,\boldsymbol{G})]^d \, d\theta < \frac{1}{2}\left[\gamma_{\pi/2}(\Psi,\xi,\boldsymbol{G})\right]^d, \tag{6.13}$$

where

$$\gamma_{\pi/2}(\Psi,\xi,\boldsymbol{G}) = \frac{1}{2Nm_c 2^{2Nm_c}} \sum_{\boldsymbol{s}\in\Psi}\sum_{k=1}^{2Nm_c}\left[\prod_{i=1}^{N}\Delta_{\pi/2,i}(\boldsymbol{s},\boldsymbol{p})\right]. \tag{6.14}$$

In (6.14), $\Delta_{\pi/2,i}(\boldsymbol{s},\boldsymbol{p})$ is calculated as $\Delta_{\theta,i}(\boldsymbol{s},\boldsymbol{p})$ by substituting $\theta = \pi/2$.

Let $\boldsymbol{\epsilon} = \boldsymbol{s} - \boldsymbol{p}$, $\boldsymbol{G}^\top = [\boldsymbol{G}_1^\top,\ldots,\boldsymbol{G}_N^\top]$, and the $2\times 2N$ sub-matrix $\boldsymbol{G}_i^\top = [\boldsymbol{g}_{1,i},\boldsymbol{g}_{2,i}]$. Hence, $\boldsymbol{G}_i$ spans over the i-th frame ($1 \le i \le N$) and $\boldsymbol{g}_{l,i} = [g_{l,1}^{(i)},\ldots,g_{l,2N}^{(i)}]^\top$ is of size $2N\times 1$ ($l = 1,2$). Following a similar approach as in [5], $\Delta_{\pi/2,i}(\boldsymbol{s},\boldsymbol{p})$ in (6.14) can be approximated at high transmitted powers as

$$\Delta_{\pi/2,i}(\boldsymbol{s},\boldsymbol{p}) = \frac{16N_0^2\eta_i}{|u_{1,i}|^4}\cdot P_s^{-2}\ln(P_s) + O(P_s^{-2}) \approx \frac{16N_0^2\|\boldsymbol{g}_{1,i}\|^2}{|\boldsymbol{g}_{1,i}{}^\top\boldsymbol{\epsilon}|^4}\cdot P_s^{-2}\ln(P_s),$$

where recall that $u_{1,i} = \boldsymbol{g}_{1,i}{}^\top\boldsymbol{\epsilon}$ and $\eta_i = \|\boldsymbol{g}_{1,i}\|^2$. Consequently, $\gamma_{\pi/2}(\Psi,\xi,\boldsymbol{G})$ in (6.14) can be also approximated as

$$\gamma_{\pi/2}(\Psi, \xi, \boldsymbol{G}) \approx \frac{16^N N_0^{2N} \cdot \mathscr{F}(\boldsymbol{G}, \boldsymbol{\epsilon})}{2N m_c 2^{2N m_c}} \cdot P_s^{-2N} \ln^N(P_s), \tag{6.15}$$

with $\mathscr{F}(\boldsymbol{G}, \boldsymbol{\epsilon})$ given by

$$\mathscr{F}(\boldsymbol{G}, \boldsymbol{\epsilon}) = \sum_{\boldsymbol{s}\in\Psi} \sum_{k=1}^{2Nm_c} \prod_{i=1}^{N} \frac{\|\boldsymbol{g}_{1,i}\|^2}{|\boldsymbol{g}_{1,i}{}^{\top}\boldsymbol{\epsilon}|^4}. \tag{6.16}$$

The parameter $\mathscr{F}(\boldsymbol{G}, \boldsymbol{\epsilon})^{-1}$ is defined as the coding gain function of the system. Now, by considering only the first term in (6.4) which is dominant at high powers, P_b can be approximated from (6.13) and (6.15) as

$$P_b \approx \frac{c_{d_H}}{2k_c} \left(\frac{16^N N_0^{2N} \cdot \mathscr{F}(\boldsymbol{G}, \boldsymbol{\epsilon})}{2N m_c 2^{2N m_c}} \right)^{d_H} \times \left[P_s^{-2} \ln(P_s) \right]^{N \cdot d_H}. \tag{6.17}$$

The term $\left[P_s^{-2}\ln(P_s)\right]^{N\cdot d_H}$ is designated as the diversity gain function of the system. From (6.17), it can be seen that as long as $\mathscr{F}(\boldsymbol{G}, \boldsymbol{\epsilon})^{-1} \neq 0$, a diversity gain function of $\left[P_s^{-2}\ln(P_s)\right]^{N\cdot d_H}$ can be fully achieved. The diversity function of the proposed system is simply $(N \cdot d_H)$-th and N-th power of the diversity gain functions of uncoded cooperative systems in [5, 6] and the NAF-BICM-ID systems in [10], respectively.

For the special case in which the mapping ξ is implemented *independently* and *identically* for each component $s_k \in \Omega$, only one element in $\boldsymbol{\epsilon}$ has a non-zero value. Hence, for every term in the $\gamma_{\pi/2}(\Psi, \xi, \boldsymbol{G})$ summation, $u_{1,i} = \epsilon^{(k)} g_{1,k}^{(i)}$ and $u_{2,i} = \epsilon^{(k)} g_{2,k}^{(i)}$ for some $1 \leq k \leq 2N$. Therefore, $\gamma_{\pi/2}(\Psi, \xi, \boldsymbol{G})$ can be simplified by averaging over the 1-bit neighbors $s, p \in \Omega$ rather than $\boldsymbol{s}, \boldsymbol{p} \in \Psi$ as

$$\gamma_{\pi/2}(\Psi, \xi, \boldsymbol{G}) = \frac{1}{2N m_c 2^{m_c}} \sum_{s\in\Omega} \sum_{k=1}^{m_c} \sum_{l=1}^{2N} \left[\prod_{i=1}^{N} \Delta_{\pi/2,i}(\boldsymbol{s}, \boldsymbol{p}) \right], \tag{6.18}$$

where now

$$\Delta_{\pi/2,i}(\boldsymbol{s}, \boldsymbol{p}) \approx \frac{16 N_0^2 \|\boldsymbol{g}_{1,i}\|^2}{|g_{1,l}^{(i)}|^4 |\epsilon|^4} \cdot P_s^{-2} \ln(P_s),$$

and $\epsilon = s - p$. Note from (6.18) that the performance of the system with independent mapping depends only on the magnitude of each component of $\boldsymbol{G}$ and not on the actual $\boldsymbol{G}$. The probability of error P_b can now be approximated as

$$P_{\mathrm{b}} \approx \frac{c_{d_{\mathrm{H}}}}{2k_{\mathrm{c}}} \left(\frac{16^N N_0^{2N} \cdot \mathscr{F}_{\mathrm{ind}}(\boldsymbol{G})}{2N m_{\mathrm{c}} 2^{m_{\mathrm{c}}}} \cdot \sum_{s \in \Omega} \sum_{k=1}^{m_{\mathrm{c}}} \frac{1}{|\epsilon|^{4N}} \right)^{d_{\mathrm{H}}} \times \left[P_s^{-2} \ln(P_s) \right]^{N \cdot d_{\mathrm{H}}}, \tag{6.19}$$

where $\mathscr{F}_{\mathrm{ind}}(\boldsymbol{G})$ becomes

$$\mathscr{F}_{\mathrm{ind}}(\boldsymbol{G}) = \left(\prod_{i=1}^{N} \left[\sum_{m=1}^{2N} |g_{1,m}^{(i)}|^2 \right] \right) \left(\sum_{l=1}^{2N} \left[\prod_{i=1}^{N} \frac{1}{|g_{1,l}^{(i)}|^4} \right] \right). \tag{6.20}$$

In this case, a full diversity gain function can be achieved as long as $\mathscr{F}_{\mathrm{ind}}(\boldsymbol{G})^{-1} \neq 0$.

6.3.2 Optimal Class of Precoders

When both diversity and coding gain functions are optimized, the probability of error can be minimized. For the general system using multi-D mapping ξ, this can be achieved by solving the following optimization problem:

$$\min_{\boldsymbol{G}} \quad \mathscr{F}(\boldsymbol{G}, \epsilon) \quad \text{s.t.} \quad \mathrm{tr}(\boldsymbol{G}\boldsymbol{G}^{\dagger}) \leq 2N. \tag{6.21}$$

It can be seen from (6.16) that a full diversity of $\left[P_s^{-2} \ln(P_s) \right]^{N \cdot d_{\mathrm{H}}}$ is guaranteed as long as $\boldsymbol{g}_{1,i}{}^{\mathsf{T}} \boldsymbol{\epsilon} \neq 0$ for all neighbors $\boldsymbol{s}, \boldsymbol{p} \in \Psi$. Hence, the rows of any optimal precoder $\boldsymbol{g}_{1,i}$ $(1 \leq i \leq N)$ must satisfy this. Given that this property is independent of the magnitude of $\boldsymbol{g}_{1,i}$, the problem in (6.21) can be carried in two steps.

First, to find the optimal power distribution η_i, let $\boldsymbol{g}_{1,i} = \sqrt{\eta_i} \tilde{\boldsymbol{g}}_i$ where $\tilde{\boldsymbol{g}}_i$ is a unit vector ($\|\tilde{\boldsymbol{g}}_i\| = 1$). The function $\mathscr{F}(\boldsymbol{G}, \epsilon)$ in (6.16) can then be re-written as

$$\mathscr{F}(\boldsymbol{G}, \boldsymbol{\epsilon}) = \sum_{s \in \Psi} \sum_{k=1}^{2N m_{\mathrm{c}}} \prod_{i=1}^{N} \frac{1}{\eta_i |\tilde{\boldsymbol{g}}_i^{\mathsf{T}} \boldsymbol{\epsilon}|^4}.$$

From the inequality of the geometric and arithmetic means and from the power constraint of the precoder, it can be shown that

$$\mathscr{F}(\boldsymbol{G}, \boldsymbol{\epsilon}) \geq \frac{1}{2^N} \sum_{s \in \Psi} \sum_{k=1}^{2N m_{\mathrm{c}}} \prod_{i=1}^{N} \frac{1}{|\tilde{\boldsymbol{g}}_i^{\mathsf{T}} \boldsymbol{\epsilon}|^4},$$

where the equality is achieved when $\eta_i = 2\ \forall i$. The objective function in (6.21) is then minimized when the $2N$ power of $\boldsymbol{G}$ is equally shared to the N rows $\boldsymbol{g}_{1,i}$, and no power is given to $\boldsymbol{g}_{2,i}$, i.e., $\boldsymbol{g}_{2,i} = \boldsymbol{0}$. The asymptotic optimal class of precoders $\boldsymbol{G}^{\star}$ for any mapping can then be written in general form as

$$\boldsymbol{G}^{\star} = \begin{pmatrix} \boldsymbol{G}_1^{\star} \\ \vdots \\ \boldsymbol{G}_N^{\star} \end{pmatrix}, \quad \text{where} \quad \boldsymbol{G}_i^{\star} = \begin{pmatrix} \boldsymbol{g}_{1,i}^{\top} \\ \boldsymbol{0} \end{pmatrix}, \tag{6.22}$$

and $\|\boldsymbol{g}_{1,i}\|^2 = 2$. Similar to [10], (6.22) indicates that the source and relay must transmit orthogonally to achieve the best asymptotic performance. That is, the source should equally distribute all its power in N broadcasting phases to transmit a superposition of $2N$ symbols, while being silent in the respective N cooperative phases.

The second step is to maximize the coding gain over the optimal class in (6.22). From this optimal class, the optimization problem in (6.21) simplifies to

$$\min_{\tilde{\boldsymbol{g}}_i} \quad \sum_{s\in\Psi} \sum_{k=1}^{2Nm_c} \prod_{i=1}^{N} \frac{1}{|\tilde{\boldsymbol{g}}_i^{\top}\boldsymbol{\epsilon}|^4} \quad \text{s.t.} \quad \|\tilde{\boldsymbol{g}}_i\| = 1. \tag{6.23}$$

Intuitively, we must find the N vectors $\tilde{\boldsymbol{g}}_i$ lying on the surface of a $2N$-D unit-radius hypersphere that minimize the objective function in (6.23). For full diversity, these vectors must not be perpendicular to any of the 1-bit error vectors $\boldsymbol{\epsilon}$, i.e., $\boldsymbol{g}_{1,i}{}^{\top}\boldsymbol{\epsilon} \neq 0$. Hence, the problem in (6.23) is constellation and mapping dependent. Furthermore, due to the number of variables involved, (6.23) is numerically and analytically very complex to solve for multi-D mappings with $N > 1$.

For the system using independent mapping ξ, the problem in (6.21) simplifies to

$$\min_{\boldsymbol{G}} \quad \mathscr{F}_{\text{ind}}(\boldsymbol{G}) \quad \text{s.t.} \quad \|\boldsymbol{G}\|^2 \leq 2N, \tag{6.24}$$

and full diversity is guaranteed as long as $\boldsymbol{g}_{1,i}$ has no zero components. By the inequality of the geometric and arithmetic means and from the power constraint of the precoder $\boldsymbol{G}$, it can be shown that

$$\mathscr{F}_{\text{ind}}(\boldsymbol{G}) \geq \left(2^N N^N \prod_{i=1}^{N}\prod_{m=1}^{2N} |g_{1,m}^{(i)}|^{\frac{1}{N}}\right)\left(2N \prod_{l=1}^{2N}\prod_{i=1}^{N} |g_{1,l}^{(i)}|^{-\frac{2}{N}}\right) \geq 2^{N+1}N^{2N+1},$$

where the equalities are achieved when $|g_{1,1}^{(i)}|^2 = |g_{1,2}^{(i)}|^2 = \cdots = |g_{1,2N}^{(i)}|^2 = 1/N$ for all $1 \leq i \leq N$. Hence, the optimization problem in (6.24) reaches its minimum when $|g_{1,k}^{(i)}|^2 = 1/N$ and $|g_{2,k}^{(i)}|^2 = 0, \forall i, k$. The asymptotically optimal precoder for independent mappings can then be written as

$$\boldsymbol{G}(\boldsymbol{\varphi}) = \frac{1}{\sqrt{N}} \begin{pmatrix} \boldsymbol{G}_1(\varphi_1) \\ \vdots \\ \boldsymbol{G}_N(\varphi_N) \end{pmatrix}, \quad \text{where} \quad \boldsymbol{G}_i(\varphi_i) = \begin{pmatrix} \boldsymbol{g}_{\varphi_i}^{\top} \\ \boldsymbol{0} \end{pmatrix}, \tag{6.25}$$

with $\boldsymbol{\varphi}^{\mathsf{T}} = [\boldsymbol{\varphi}_1^{\mathsf{T}}, \ldots, \boldsymbol{\varphi}_N^{\mathsf{T}}]$, $\boldsymbol{\varphi}_i = [\varphi_1^{(i)}, \ldots, \varphi_{2N-1}^{(i)}]^{\mathsf{T}}$ and $\boldsymbol{g}_{\boldsymbol{\varphi}_i}^{\mathsf{T}} = [1, e^{j\varphi_1^{(i)}}, \cdots, e^{j\varphi_{2N-1}^{(i)}}]$. In the next subsection, we shall further address the design of the angles $\boldsymbol{\varphi}$ to achieve good convergence properties for independent mappings.

6.3.3 Design of Superposition Angles for Independent Mappings

Among a wide range of optimal $\boldsymbol{G}(\boldsymbol{\varphi})$ in (6.25), the one that yields a faster convergence to the asymptotic performance is preferred. Therefore, one also needs to take into account the performance at the first iteration. A reasonable approach is to consider the worst-case of PEP in (6.7) and to find $\boldsymbol{\varphi}$ that minimizes this PEP.

Applying the Chernoff bound to the Gaussian integral in (6.7), the PEP can be approximated at high powers as

$$\mathbb{P}(\underline{\boldsymbol{s}} \to \underline{\breve{\boldsymbol{s}}}) \approx \frac{1}{2} \prod_{e=1}^{d} \prod_{i=1}^{N} \Delta_{\pi/2,i}\left(\boldsymbol{s}_{(e)}, \breve{\boldsymbol{s}}_{(e)}\right). \tag{6.26}$$

Using the $\boldsymbol{G}(\boldsymbol{\varphi})$ precoder and following a similar approach as in the previous section, the PEP can be further approximated as

$$\mathbb{P}(\underline{\boldsymbol{s}} \to \underline{\breve{\boldsymbol{s}}}) \approx \frac{32^{dN} N_0^{2dN} P_s^{-2dN} \ln^{dN}(P_s)}{2} \prod_{e=1}^{d} \prod_{i=1}^{N} \frac{1}{|u_{1,i}^{(e)}|^4}, \tag{6.27}$$

where $u_{1,i}^{(e)} = \boldsymbol{g}_{\boldsymbol{\varphi}_i}^{\mathsf{T}} \boldsymbol{\epsilon}_{(e)} = \epsilon_1^{(e)} + \sum_{k=1}^{2N-1} \epsilon_{k+1}^{(e)} \exp(j\varphi_k^{(i)})$. Note from (6.27) that the worst-case PEP is independent of e and hence the super-frame index can be dropped. The worst-case PEP can then be minimized by solving the following:

$$\max_{\boldsymbol{\varphi}} \min_{\substack{\boldsymbol{s}, \breve{\boldsymbol{s}} \in \Psi, \\ \boldsymbol{s} \neq \breve{\boldsymbol{s}}}} \prod_{i=1}^{N} |u_{1,i}|^4, \tag{6.28}$$

where $u_{1,i} = \boldsymbol{g}_{\boldsymbol{\varphi}_i}^{\mathsf{T}} \boldsymbol{\epsilon} = \epsilon_1 + \sum_{k=1}^{2N-1} \epsilon_{k+1} \exp(j\varphi_k^{(i)})$, with $\boldsymbol{\epsilon} = \boldsymbol{s} - \breve{\boldsymbol{s}} = (\epsilon_1, \epsilon_2, \ldots, \epsilon_{2N})^{\mathsf{T}}$.

The optimization problem in (6.28) is constellation dependent and does not have a general solution. For the $N = 1$, this problem has been analytically solved in [10] for several M-QAM schemes. However, for $N > 1$, the problem in (6.28) is mathematically and numerically involved. As an alternative, we propose a pragmatic approach based on (6.28) to find good rotation angles $\boldsymbol{\varphi}$.

First, the optimization problem in (6.28) can be relaxed to find the optimal $\boldsymbol{\varphi}_1^{\star}$ as

$$\max_{\varphi_1} \min_{\substack{s,\breve{s}\in\Psi,\\ s\neq\breve{s}}} |\mathbf{g}_{\varphi_1}^{\top}\boldsymbol{\epsilon}|^4. \tag{6.29}$$

This guarantees that the first term of the product in (6.28) is reasonably large. The solution to (6.29) is similar to that of multi-user systems in [4] solved numerically for various values of N and several square M-QAM modulation schemes. Then, to find $\boldsymbol{\varphi}_i$ for $2 \leq i \leq N$, we use the vector $\mathbf{g}_{\varphi_1^*}$ obtained from (6.29) for all i, except that the signs of N values in this vector are flipped to make sure that all vectors in the set $\{\mathbf{g}_{\varphi_i}\}$ are orthogonal. The product in (6.28) is then reasonably large. For instance, by applying this approach to the case of $N = 2$, one has

$$\mathbf{G}(\boldsymbol{\varphi}) = \frac{1}{\sqrt{2}} \begin{pmatrix} 1 & e^{j\varphi_1} & e^{j\varphi_2} & e^{j\varphi_3} \\ 0 & 0 & 0 & 0 \\ 1 & -e^{j\varphi_1} & e^{j\varphi_2} & -e^{j\varphi_3} \\ 0 & 0 & 0 & 0 \end{pmatrix}. \tag{6.30}$$

In (6.30), the angles $\{\varphi_1, \varphi_2, \varphi_3\}$ are the solution to (6.29) and are provided in [4]. As will be shown in the Sect. 6.5, the system using the optimal precoder in (6.30) presents significant coding gains as compared to other precoders.

6.4 Near-Capacity Design in Turbo Pinch-Off Areas

The diversity analysis presented earlier is useful in the BER floor region. In this region, a low BER can be achieved at a sufficiently high transmitted powers. Also of importance is the so-called turbo pinch-off or waterfall region, where a significant BER reduction is observed over iterations. The turbo pinch-off region can be used to examine whether our coded system can achieve near-capacity performance.

In order to approach near-capacity performance, state-of-the-art codes such as turbo and turbo-like codes are usually used. In fact, such powerful codes have been successfully applied to many channels, including the relay channels in [9, 17, 18]. Lately, over Rayleigh fading channels, reference [8] proposes a powerful coding scheme in which a multi-D mapping is employed in a rotated mutli-D constellation. Interestingly, it is demonstrated in [8] that by only using a simple outer convolutional code, the proposed technique outperforms any coded system using turbo-like codes with traditional modulation schemes.

In this section, using EXIT chart analysis [12], we show that by employing a suitable multi-D mapping ξ in multiple precoded cooperative frames, the considered BICM-ID system can also offer a capacity-approaching performance. For simplicity, particular attention is paid to the case of $N = 2$.

6.4.1 Multi-D Mapping in Precoded Multiple Cooperative Frames

Here, we propose the use of multi-D mapping ξ in N precoded cooperative frames along with a full-diversity rotation $\boldsymbol{G}$ [3]. Full diversity rotations have the property that $\boldsymbol{x} - \breve{\boldsymbol{x}} = \boldsymbol{G}(\boldsymbol{s} - \breve{\boldsymbol{s}})$ contains all nonzero components as long as $\boldsymbol{s} \neq \breve{\boldsymbol{s}}$. Note from (6.17) that this type of rotation achieves the full cooperative diversity $\left[P_s^{-2}\ln(P_s)\right]^{N \cdot d_{\mathrm{H}}}$ of the NAF-BICM system for any labeling technique. Furthermore, full-diversity rotations can increase the achievable rate over ergodic fading channels [8]. For example, Fig. 6.2 shows the rates over a NAF channel achieved by three different inputs, namely Gaussian input, QPSK constellation, and 4-D precoded QPSK constellation using the 4×4 full-diversity rotation $\boldsymbol{G}_{\mathrm{st}}$ given in [3]:

$$\boldsymbol{G}_{\mathrm{st}} = \frac{1}{\sqrt{4}} \begin{pmatrix} 1 & \omega_1 & \omega_1^2 & \omega_1^3 \\ 1 & \omega_2 & \omega_2^2 & \omega_2^3 \\ 1 & \omega_3 & \omega_3^2 & \omega_3^3 \\ 1 & \omega_4 & \omega_4^2 & \omega_4^3 \end{pmatrix}, \tag{6.31}$$

with $\omega_1 = \omega = \exp(j\pi/8)$ and $\omega_i = \omega \exp(j\pi(i-1)/2)$. As can be seen in Fig. 6.2, the rate can be greatly improved using the rotated multi-D constellation.

For a good mapping rule ξ, one must minimize $\gamma_{\pi/2}(\Psi, \xi, \boldsymbol{G})$ in (6.14). To obtain more insights about the mapping rule, we first approximate $\Delta_{\pi/2,i}(\boldsymbol{s}, \boldsymbol{p})$ in (6.14) as

$$\Delta_{\pi/2,i}(\boldsymbol{s}, \boldsymbol{p}) \approx \frac{1}{(c_{\pi/2}|u_{1,i}|^2 + 1)^2} \left\{ 1 + \left(\frac{1}{P_s b^2} - a \right) \ln \left(1 + \frac{1}{a} \right) \right\}. \tag{6.32}$$

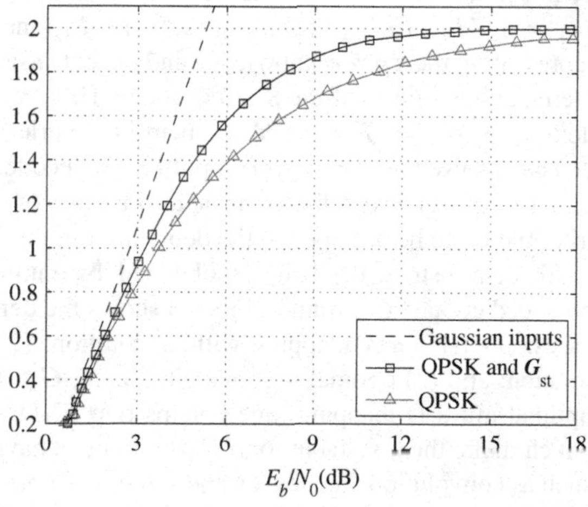

Fig. 6.2 Achievable rates over NAF channels with different inputs

Observe that $\Delta_{\pi/2,i}(\boldsymbol{s}, \boldsymbol{p})$ is only a function of $(|u_{1,i}|^2, |u_{2,i}|^2)$, where recall that $\boldsymbol{u}_i = [u_{1,i}^{\top}, u_{2,i}^{\top}]^{\top} = \boldsymbol{G}_i(\boldsymbol{s} - \boldsymbol{p})$. Hence, $\Delta_{\pi/2,i}(\boldsymbol{s}, \boldsymbol{p})$ can be rewritten as $\Delta_{\pi/2,i}(|u_{1,i}|^2, |u_{2,i}|^2)$. As shown in [11], $\Delta_{\pi/2,i}(\boldsymbol{s}, \boldsymbol{p})$ is a strictly decreasing function of $|u_{1,i}|^2$ and $|u_{2,i}|^2$. Given the fact that $|u_{1,i}|^2$ and $|u_{2,i}|^2$ are both positive for a full diversity $\boldsymbol{G}$, it can then be seen that the BER can be minimized by maximizing $|u_{1,i}|^2$ and $|u_{2,i}|^2$ for each term in the summation (6.14). Since $\boldsymbol{G}$ is also unitary, it is equivalent to maximizing $\|\boldsymbol{u}\|^2 = \|\boldsymbol{G}(\boldsymbol{s} - \boldsymbol{p})\|^2 = \|\boldsymbol{G}\epsilon\|^2$ for all 1-bit neighbors $\boldsymbol{s}$ and $\boldsymbol{p}$. Using (6.14), it turns out that the design criterion is to maximize the average Euclidean distance between all 1-bit neighbors in the multi-D constellation Ψ. Interestingly, such a design criterion for multi-D mapping is the same as that over a single-antenna additive white Gaussian noise (AWGN) channel in [14]. Therefore, the optimal multi-D mapping for NAF channels can be obtained in a similar manner. As shall be seen shortly, by using such a multi-D mapping, one can select a relatively simple outer code for capacity-approaching performance.

6.4.2 EXIT Chart Analysis

To demonstrate the advantage of the proposed coding scheme, we first apply the EXIT chart technique [12] to examine the demapper's EXIT curves. Then a combination of the demapper and a simple convolutional decoder, with which close-capacity performance can be achieved, is studied. Due to space limitation, we only consider the QPSK modulation scheme and adopt the optimal 4-D mapping proposed in [14].

Similar to [12], let I_{A_1} and I_{E_1} denote the mutual information between the *a priori* log-likelihood ratio (LLR) and the transmitted coded bit, and between the extrinsic LLR and the transmitted coded bit at the input and output of the demodulator, respectively. The demodulator EXIT characteristic is then given by $I_{E_1} = T_1(I_{A_1}, E_b/N_0)$ [12]. Similarly, let I_{A_2} and I_{E_2} be the mutual information representing the *a priori* knowledge and the extrinsic information of the coded bits at the input and output of the SISO decoder. The decoder EXIT characteristic is then defined as $I_{E_2} = T_2(I_{A_2})$. After being deinterleaved, the extrinsic output of the detector is used as the *a priori* input to the decoder, i.e., $I_{A_2} = I_{E_1}$. Furthermore, after being interleaved, the extrinsic information of the decoder becomes the *a priori* information to be provided to the demodulator, i.e., $I_{A_1} = I_{E_2}$.

To demonstrate the benefits obtained by multi-D mapping employed in two precoded cooperative frames, Fig. 6.3 shows the demmaper's EXIT curves for three cases: (1) 1-D Gray mapping without rotation; (2) optimal 4-D mapping without rotation; and (3) optimal 4-D mapping using $\boldsymbol{G}_{\mathrm{st}}$. It can be seen that by using the optimal multi-D mapping, one obtains two EXIT curves with a very steep slope, which make them suitable for a class of codes having also a decayed EXIT curve, such as convolutional codes. Furthermore, with the same optimal 4-D mapping, the full diversity rotation provides larger I_{E_1} in a wide range of I_{A_1}. This advantage comes from the capacity improvement discussed earlier.

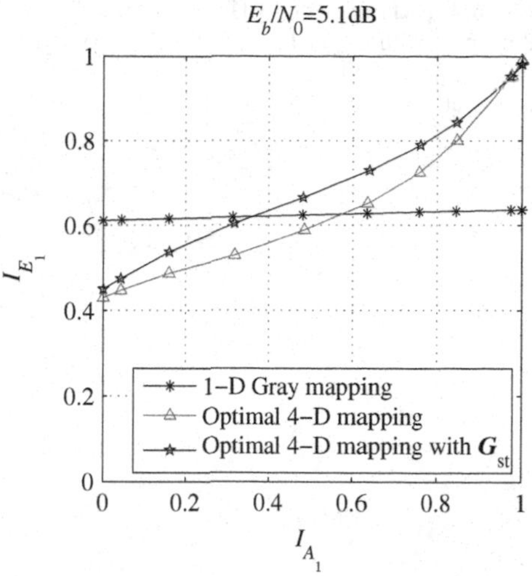

Fig. 6.3 Demapper EXIT curves for three different systems

Now, given the superiority of multi-D mapping and precoding, we can apply the EXIT chart technique [12] to select the most suitable convolutional code for the proposed system. By using EXIT charts, both EXIT curves of the multi-D demodulator and the decoder are plotted in the same graph, but the axes of the decoder curve are swapped [12]. The convergence behavior can therefore be visualized.

We have examined different code rates and observed that the multi-D demapper EXIT curve can be matched with very simple codes. For example, Fig. 6.4 plots the demapper EXIT curves at $E_b/N_0 = 5.1$ dB, along with that of the simple rate-2/3, 4-state punctured convolutional code obtained from the rate-1/2 convolutional code with generator matrix {5; 7} [1]. The above SNR is chosen to make sure that an open tunnel exists between the two curves in Fig. 6.4. Note that at this rate, the ergodic capacity of the NAF channel calculated as in [7] and the constrained capacity with QPSK are at $E_b/N_0 = 3.6$ dB and $E_b/N_0 = 5.6$ dB, respectively. It can be seen from Fig. 6.4 that the two EXIT curves match very well at the chosen SNR. Specifically, the two curves only intersect at an ending point that is very close to $I_{A_1}(1)$. Apparently, our proposed system can operate below the constrained capacity with QPSK. This fact is later confirmed by simulations.

The proposed scheme can also work well with other code rates. Though not explicitly shown here, we observe that one can achieve a good match for the rate-3/4 system at $E_b/N_0 = 5.9$ dB using the rate-3/4, 4-state punctured convolutional code obtained from the rate-1/2 convolutional code with generator matrix {5; 7} [1]. It should be mentioned that the ergodic capacity with Gaussian inputs and the constrained capacity with QPSK for this rate are at $E_b/N_0 = 4.05$ dB and $E_b/N_0 = 7$ dB, respectively

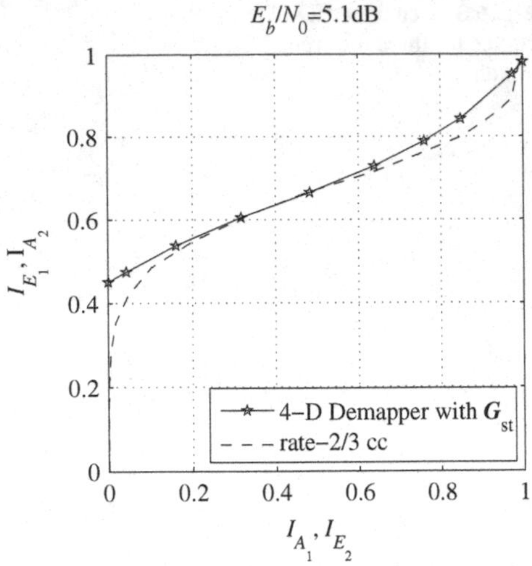

Fig. 6.4 EXIT charts of 4-D demapper with $\boldsymbol{G}_{\text{st}}$ and rate-2/3 punctured convolutional code

6.5 Illustrative Examples

In this section, analytical and simulation results are provided to confirm the tightness of the derived bound and the superiority of the multiple-frame precoding technique in both error-floor and turbo pinch-off regions. Unless otherwise stated, a random interleaver of length 80,000 bits is used in all simulations. In the computation of the bound to P_{b} in (6.4), the first twenty terms of the summation are retained to provide an accurate bound. Furthermore, for simplicity, we only consider the NAF-BICM-ID system using the QPSK modulation scheme.

To verify the tightness of the derived bound when precoding over $N = 2$ cooperative frames, Fig. 6.5 presents the BER performances after five iterations of the NAF-BICM-ID system employing three different 4×4 precoders. The considered precoders include $\boldsymbol{I}_4$, i.e., no precoder, and the 4×4 $\boldsymbol{G}_{\text{mixed}}$ and $\boldsymbol{G}_{\text{krus}}$ precoders given in [16]. Gray mapping and the rate-$1/2$ 4-state convolutional code with generator matrix $\{5; 7\}$ are considered in this and the next figure. Observe from Fig. 6.5 that the error performances of all systems converge to their respective bounds at practical BER levels. In particular, the analytical and simulation results converge around the BER level of 10^{-3} for all systems. This makes the bound an effective tool to predict the error performance of systems precoding over several frames.

To confirm the optimality of the proposed multiple-frame precoding scheme in error-floor regions, Fig. 6.6 shows the BER performances after five iterations of the systems precoding over $N = 2$ with the following 4×4 matrices: (1) $\boldsymbol{I}_4$, no precoder; (2) $\boldsymbol{G}_{\text{st}}$ in (6.31), which is the optimal precoder over single-antenna fading channels [3, 15]; (3) $\boldsymbol{G}_{\pi/6\,\text{Bk}} = \text{bdiag}\left(\boldsymbol{G}_{\pi/6}, \boldsymbol{G}_{\pi/6}\right)$ with $\boldsymbol{G}_{\pi/6} = \begin{pmatrix} 1 & e^{j\pi/6} \\ 0 & 0 \end{pmatrix}$, which is equivalent to the optimal single-frame precoder [10]; and (4) the proposed

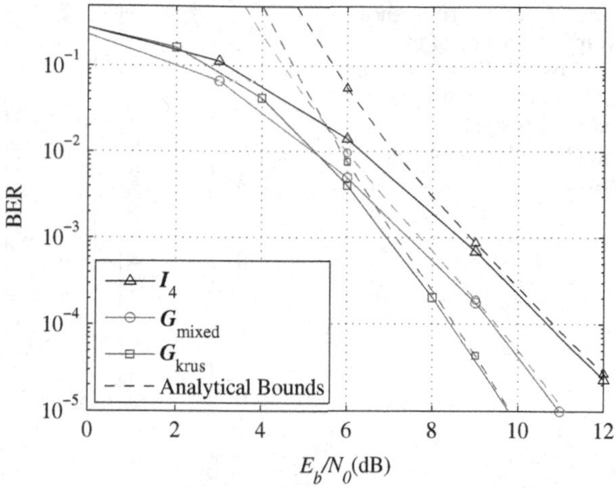

Fig. 6.5 BER performances of the NAF-BICM-ID system with different 4×4 precoders

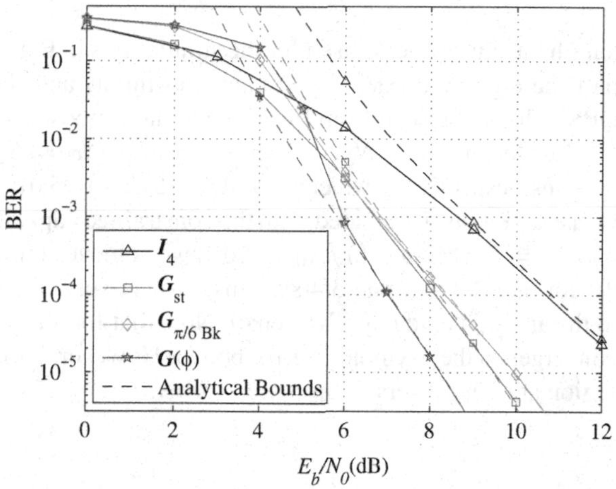

Fig. 6.6 BER performances of the NAF-BICM-ID system with $\boldsymbol{I}_4$, $\boldsymbol{G}_{\text{st}}$, $\boldsymbol{G}_{\pi/6\,\text{Bk}}$ and $\boldsymbol{G}(\varphi)$

$\boldsymbol{G}(\varphi)$ in (6.30), which belongs to the optimal class in (6.25) with the angles for good convergence behavior $\varphi_1 = 0.15$, $\varphi_2 = 0.30$ and $\varphi_3 = 0.62$ as derived in [4]. First, it can be seen from Fig. 6.6 that precoding over two frames as in the case of $\boldsymbol{G}_{\text{st}}$ and $\boldsymbol{G}(\varphi)$ provides a higher diversity order than precoding over a single frame (i.e., $\boldsymbol{G}_{\pi/6\,\text{Bk}}$) or no precoding at all. It can be easily verified from (6.19) that both $\boldsymbol{G}_{\text{st}}$ and $\boldsymbol{G}(\varphi)$ achieve full cooperative diversity. More importantly, the system using $\boldsymbol{G}(\varphi)$ greatly outperforms the systems using the other precoders. In particular, the proposed optimal precoder presents coding gains of 4.7 dB, 1.7 dB, and 1.2 dB over $\boldsymbol{I}_4$, $\boldsymbol{G}_{\pi/6\,\text{Bk}}$, and $\boldsymbol{G}_{\text{st}}$, respectively, at the BER level of 10^{-5}.

Finally, to demonstrate the advantage of multiple-frame precoding in near-capacity regions, Fig. 6.7 shows the BER performance after 50 decoding iterations of the systems using the $\boldsymbol{G}_{\text{st}}$ precoder, the multi-D mapping from [14], and the rate-2/3 and rate-3/4 punctured convolutional codes mentioned in Sect. 6.4. A 160,000-

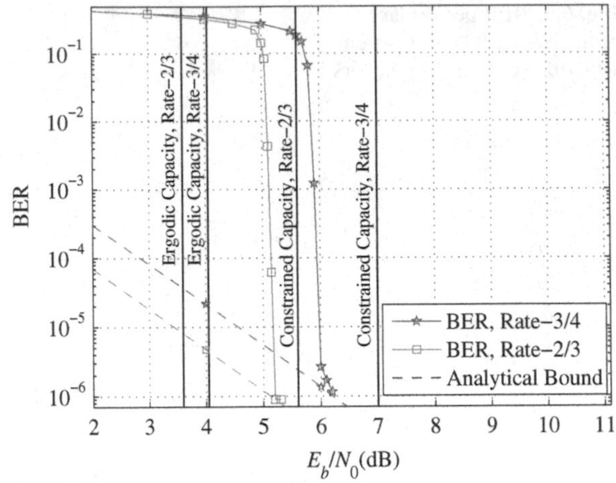

Fig. 6.7 BER performances with 50 iterations of the NAF-BICM-ID system using the $\boldsymbol{G}_{\text{st}}$ precoder, the multi-D mapping from [14], and the rate-2/3 and rate-3/4 outer codes

length bit-interleaver is used for the simulations in Fig. 6.7. For comparison, we also plot the ergodic capacity with Gaussian inputs and the constrained capacity with QPSK. It can be seen from Fig. 6.7 that the proposed system can achieve a BER of 10^{-5} or lower at $E_b/N_0 = 5.15$ dB and $E_b/N_0 = 6$ dB for rate-2/3 and rate-3/4 systems, respectively, which are only 1.55 dB–1.95 dB away from the capacity with Gaussian inputs. Compared with the constrained capacities with QPSK, which are at $E_b/N_0 = 5.6$ dB and $E_b/N_0 = 7$ dB, we achieve significant gains of 0.45 dB and 1 dB for rate-2/3 and rate-3/4 systems, respectively. These results clearly match well to the analysis made by EXIT chart. Note that for all systems, the BER performance converges to the asymptotic error bound. However, it only happens in the error-floor region at high powers as discussed earlier.

6.6 Concluding Remarks

This chapter considered a precoding scheme over multiple cooperative frames for NAF-BICM-ID systems. A tight union bound on the BER was first derived for the proposed system. It was then shown that multiple-frame precoding can provide significant benefits with respect to both diversity function in the error-floor region and capacity-approaching performance in the turbo pinch-off region. Specifically, in the error-floor region, an optimal class of precoders with respect to both diversity and coding gains was developed for independent and multi-D mapping. In the turbo pinch-off area, by concatenating the proposed system with a simple outer code and applying multi-D mapping, we demonstrated that one can approach close to the ergodic capacity with Gaussian inputs.

References

1. Begin, G., Haccoun, D., Paquin, C.: Further results on high-rate punctured convolutional codes for Viterbi and sequential decoding. IEEE Trans. Commun. **38**(11), 1922–1928 (1990). Doi:10.1109/26.61470
2. Benedetto, S., Divsalar, D., Montorsi, G., Pollara, F.: A soft-input soft-output APP module for iterative decoding of concatenated codes. IEEE Commun. Lett. **1**(1), 22–24 (1997). Doi:10.1109/4234.552145
3. Boutros, J., Viterbo, E.: Signal space diversity: A power and bandwidth efficient diversity technique for the Rayleigh fading channel. IEEE Trans. Inf. Theory **44**(4), 1453–1467 (1998). Doi:10.1109/18.681321
4. Damen, M.: Joint coding/decoding in a multiple access system, application to mobile communications. Ph.D. thesis, ENST de Paris, France (1999)
5. Ding, Y., Zhang, J.K., Wong, K.M.: The amplify-and-forward half-duplex cooperative system: Pairwise error probability and precoder design. IEEE Trans. Signal Process. **55**(2), 605–617 (2007). Doi:10.1109/TSP.2006.885761
6. Ding, Y., Zhang, J.K., Wong, K.M.: Optimal precoder for amplify-and-forward half-duplex relay system. IEEE Trans. Wirel. Commun. **7**(8), 2890–2895 (2008). Doi:10.1109/TWC.2008.070226
7. Ding, Y., Zhang, J.K., Wong, K.M.: Ergodic channel capacities for the amplify-and-forward half-duplex cooperative systems. IEEE Trans. Inf. Theory **55**(2), 713–730 (2009). Doi:10.1109/TIT.2008.2009822
8. Herath, S., Tran, N., Le-Ngoc, T.: Rotated multi-D constellations in Rayleigh fading: Mutual information improvement and pragmatic approach for near-capacity performance in high-rate regions. IEEE Trans. Commun. **60**(12), 3694–3704 (2012). Doi:10.1109/TCOMM.2012.091012.110253
9. Hu, J., Duman, T.: Low density parity check codes over wireless relay channels. IEEE Trans. Wirel. Commun. **6**(9), 3384–3394 (2007). Doi:10.1109/TWC.2007.06083
10. Jiménez-Rodríguez, L., Tran, N.H., Le-Ngoc, T.: Bandwidth-efficient bit-interleaved coded modulation over NAF relay channels: Error performance and precoder design. IEEE Trans. Veh. Technol. **60**(5), 2086–2101 (2011). Doi:10.1109/TVT.2011.2138176
11. Jiménez-Rodríguez, L., Tran, N.H., Le-Ngoc, T.: Multiple-frame precoding and multi-D mapping for BICM over ergodic NAF relay channels. Wiley J. Wirel. Commun. Mobile Comput. **11**(12), 1564–1575 (2011). Doi:10.1002/WCM.1242
12. Ten Brink, S.: Designing iterative decoding schemes with the extrinsic information chart. AEU Int. J. Electron. Commun. **54**(6), 389–398 (2000)
13. Tran, N.H., Jiménez-Rodríguez, L., Le-Ngoc, T., Bahrami, H.R.: Precoding and symbol grouping for NAF relaying in BICM systems. IEEE Trans. Veh. Technol. **62**(6), 2607–2617 (2013). Doi:10.1109/TVT.2013.2245927
14. Tran, N.H., Nguyen, H.H.: Design and performance of BICM-ID systems with hypercube constellations. IEEE Trans. Wirel. Commun. **5**(5), 1169–1179 (2006). Doi:10.1109/TWC.2006.1633370
15. Tran, N.H., Nguyen, H.H., Le-Ngoc, T.: Performance of BICM-ID with signal space diversity. IEEE Trans. Wirel. Commun. **6**(5), 1732–1742 (2007). Doi:10.1109/TWC.2007.360375
16. Viterbo, E.: Full diversity rotations. URL http://www.ecse.monash.edu.au/staff/eviterbo/rotations/rotations.html
17. Zhang, Z., Duman, T.: Capacity-approaching turbo coding and iterative decoding for relay channels. IEEE Trans. Commun. **53**(11), 1895–1905 (2005). Doi:10.1109/TCOMM.2005.858654
18. Zhang, Z., Duman, T.: Capacity-approaching turbo coding for half-duplex relaying. IEEE Trans. Commun. **55**(10), 1895–1906 (2007). Doi:10.1109/TCOMM.2007.906404

Chapter 7
Full-Duplex AF Relaying: Capacity Under Residual Self-Interference

In all previous chapters, we addressed the capacity and code designs for several HD relay protocols. We now turn our attention to FD relay schemes. As explained in Chap. 1, pioneering works on FD relaying concentrated on the ideal scenario in which the relay is able to transmit and receive simultaneously without any self-interference. Since self-interference cannot be completely mitigated in practice, this ideal assumption overestimated the benefits of FD transmission. As such, current studies on FD relaying take the effect of *residual* self-interference into account [1, 5, 10–19, 21]. In particular, by considering some imperfection in the cancellation process (e.g., imperfect channel state information or imperfect knowledge of the transmitted signal), most of these works assume that the variance of the residual self-interference is proportional to the average transmitted power [1, 5, 10, 11, 13–19, 21]. Under this assumption, the performance of FD relaying has been shown to be severely degraded in terms of rate and reliability, specially in high power regions. For instance, the achievable rate of a FD DHAF system was investigated in [16, 18] along with break-even boundaries between HD and FD modes. From the boundaries in [16, 18], HD is preferred in high relay power regions when full power allocation is applied. To control the self-interference, optimal power allocation strategies were then derived in [16, 18], where it was observed that full power at the FD relay is not necessarily optimal.

Recently, experimental results in [6] suggest that the previous assumption on the variance of the interference might only correspond to the worst-case scenario. Specifically, it was observed in [6] that the variance of the residual self-interference after analog and/or digital cancellation is proportional to the λ-th power of the average transmitted power ($0 \leq \lambda \leq 1$), where λ indicates the quality of the self-interference cancellation technique in use. Similar trends for other mitigation prototypes can also be observed in [3]. Different from previous theoretical works that focused on $\lambda = 1$, the experimental results in [3, 6] indicate that λ is often less than one. The main reason for this behavior is that it is easier to estimate the

L. Jiménez Rodríguez et al., *Amplify-and-Forward Relaying in Wireless Communications*, SpringerBriefs in Computer Science,
DOI 10.1007/978-3-319-17981-0_7

self-interference as its power increases. Therefore, more benefits of FD relaying can be expected under this empirical model. In addition, it is clear that the optimal allocations derived earlier in [16, 18] no longer hold true.

This chapter investigates the capacity and respective optimal power allocation strategy for a FD DH system without direct link and under the residual self-interference model in [6]. The focus is on the static scenario similar to Chap. 3. Both individual and global power constraints are considered. First, the optimal power allocation schemes that maximize the achievable rates are derived. The capacity and optimal power strategies are then investigated in different high power regions. Specifically, we apply the method of dominant balance [2, Ch. 3.4] to show that using full power at the relay might be detrimental to the achievable rate. Following a similar approach, the multiplexing gain of the FD system using the optimal power allocation is shown to be $1/[1+\lambda]$. The FD DH scheme is finally compared to the HD system. Analytical and simulation results reveal that FD is advantageous when the source has a large power constraint and the relay does not, or when both the source and relay have a large constraint and $\lambda < 1$.

7.1 Problem Formulation

In this chapter, we consider the static scenario similar to Chap. 3. In this scenario, the channel gains $\boldsymbol{h} = [h_1, h_2]$ are not time-varying so that full channel knowledge can be acquired at all nodes. Since full CSI is available, the CI coefficient $b = \sqrt{z_2/[\alpha_1 q_1 P_s + N_r + V]}$ in (2.25) is used. The FD DHAF system with input-output relation in (2.24) is adopted. In particular, we assume that the direct source-destination link is under deep shadowing so that $h_0 = 0$ in (2.24).

The distribution of the self-interference after several stages of cancellation is unknown in practice. Here, similar to previous works, we assume that the residual self-interference is zero-mean, additive and Gaussian as $\mathscr{CN}(0, V)$. The Gaussian assumption might hold in reality due to the various sources of imperfections in the cancellation process (i.e., due to the central limit theorem). If the interference is not Gaussian, this assumption can be considered as the worst-case scenario in terms of achievable rate [20]. In addition, based on the experimental results in [6], the variance of the residual self-interference is modeled as

$$V = \beta(z_2 P_r)^{\lambda}, \tag{7.1}$$

where $z_2 P_r$ is the average power transmitted by the relay; and β (given in units of $[\text{Watts}]^{1-\lambda}$) and λ ($0 \leq \lambda \leq 1$) are constants that reflect the quality of the selected cancellation technique [3, 6]. Note that the model in (7.1) includes two important scenarios: the optimistic case in which the interference variance is simply a constant and not a function of transmitted power ($\lambda = 0$) [3, 7], and the pessimistic one in which the variance increases linearly with transmitted power ($\lambda = 1$) [1, 10, 11, 13–19, 21]. As we will show in this and the following chapter, the value of λ plays a crucial role in the performance of FD relay systems.

Substituting the amplification coefficient b and the self-interference V, the achievable rate in (2.26) can be written when $\alpha_0 = 0$ as

$$I_{\mathrm{FD}}|\boldsymbol{h} = \log\left(1 + \frac{q_1 z_2 \gamma_1 \gamma_2}{q_1\gamma_1 + z_2\gamma_2 + z_2^{\lambda}\gamma_3 + z_2^{1+\lambda}\gamma_2\gamma_3 + 1}\right) = \log[1 + f(q_1, z_2)], \tag{7.2}$$

where $\gamma_1 = [P_s\alpha_1]/N_r$, $\gamma_2 = [P_r\alpha_2]/N_d$, and $\gamma_3 = [\beta P_r^{\lambda}]/N_r$. The objective of this chapter is to maximize the above rate. First, in the individual constraint scenario, we assume that $q_1 \leq q_s$ and $z_2 \leq z_r$. From (2.24), the power constraints at S and R are $q_s P_s$ and $z_r P_r$, respectively. The constants q_s and z_r might be set to one to have constraints of P_s and P_r. Similar to [9, 18], we also consider the global constraint scenario with $P_s = P_r = P_t$ and $q_1 + z_2 \leq q_t$, where $q_t P_t$ is the joint power constraint. The capacity of the FD DH system can be obtained by solving the following:

$$C_{\mathrm{FD}} = \max_{q_1, z_2 \geq 0} \quad I_{\mathrm{FD}}|\boldsymbol{h}, \quad \text{s.t.} \begin{cases} q_1 \leq q_s, z_2 \leq z_r & \text{(indiv.)} \\ q_1 + z_2 \leq q_t & \text{(joint).} \end{cases} \tag{7.3}$$

The rate in (7.2) is in general not a concave function of $\boldsymbol{q} = [q_1, z_2]$. However, the problem in (7.3) is quasiconcave and has a unique global maximizer $\boldsymbol{q}^\star = [q_1^\star, z_2^\star]$, as detailed in the next section.

7.2 Capacity and Optimal Power Allocation

In this section, we derive the capacity in (7.3) along with the optimal power allocation schemes. Since $\log(\cdot)$ is monotonically increasing, (7.3) can be equivalently solved by maximizing $f(\cdot,\cdot)$ in (7.2).

Proposition 7.1. *The optimal allocation under individual constraints is given by*

$$q_1^\star = q_s, \qquad z_2^\star = \begin{cases} z_r, & \lambda = 0 \\ \min\{p_1, z_r\}, & 0 < \lambda < 1 \\ \min\left\{\sqrt{\frac{q_s\gamma_1 + 1}{\gamma_2\gamma_3}}, z_r\right\}, & \lambda = 1, \end{cases} \tag{7.4}$$

where p_1 is the only positive root of $P_1(z_2, \lambda) = a_1 z_2^{1+\lambda} + b_1 z_2^{\lambda} + c_1$, with $a_1 = -\lambda\gamma_2\gamma_3$, $b_1 = \gamma_3(1-\lambda)$ and $c_1 = q_s\gamma_1 + 1$.

Proof. This follows from the fact that $f(\cdot)$ is increasing with q_1 for any $0 \leq \lambda \leq 1$, increasing with z_2 for $\lambda = 0$, and quasiconcave [4, Ch.3.4] in z_2 for $\lambda > 0$ [8]. □

Proposition 7.2. *The optimal power allocation under global constraints is given by*

$$z_2^\star = q_t - q_1^\star, \qquad q_1^\star = \begin{cases} \frac{-B_0-\sqrt{B_0^2-4A_0C_0}}{2A_0}, & \lambda = 0, A_0 \neq 0 \\ q_t/2, & \lambda = 0, A_0 = 0 \\ p_2, & 0 < \lambda < 1 \\ \frac{-B_1-\sqrt{B_1^2-4A_1C_1}}{2A_1}, & \lambda = 1, A_1 \neq 0 \\ q_t/2, & \lambda = 1, A_1 = 0, \end{cases} \tag{7.5}$$

where $A_0 = \gamma_2 - \gamma_1 + \gamma_2\gamma_3$, $B_0 = -2(q_t\gamma_2+1)(\gamma_3+1)$, $C_0 = q_t(\gamma_3+1)(q_t\gamma_2+1)$, $A_1 = \gamma_2-\gamma_1+\gamma_3+q_t\gamma_2\gamma_3$, $B_1 = -2(q_t\gamma_3+1)(q_t\gamma_2+1)$, $C_1 = q_t(q_t\gamma_3+1)(q_t\gamma_2+1)$*, and* p_2 *is the only positive root of* $P_2(q_1,\lambda) = q_t - 2q_1 - q_1^2\gamma_1 + \gamma_2(q_1-q_t)^2 + \gamma_3(q_t-q_1)^\lambda[q_t - 2q_1 + \gamma_2(q_1-q_t)^2 + \lambda q_1(q_t\gamma_2 - q_1\gamma_2 + 1)]$.

Proof. This follows from the fact that full power $q_1 + z_2 = q_t$ is optimal and $f(q_1, q_t - q_1)$ is quasiconcave with respect to q_1 [8]. □

Note from the above propositions that $P_1(z_2,\lambda)$ and $P_2(q_1,\lambda)$ are highly nonlinear when $0 < \lambda < 1$. Thus, p_1 and p_2 cannot be obtained in closed-form and bisection search is needed to find the optimal allocations when $0 < \lambda < 1$.

7.3 Asymptotic Analysis

In the previous section, we derived optimal power allocation strategies that maximize the achievable rate in (7.2) under individual and global power constraints. In this section, we provide further insights on the behavior of the capacity in (7.3) and the optimal schemes in (7.4) and (7.5) in different asymptotically high power regions. Specifically, we consider the following asymptotic cases.

7.3.1 *Large Source Power, Fixed Relay Power*

Consider first the case under individual constraints where $P_s \to \infty$ while P_r remains fixed. When $0 < \lambda < 1$, applying the method of dominant balance [2, Ch. 3.4] to $P_1(z_2,\lambda)$, it can be shown that $O(z_2^{1+\lambda}) = O(q_s\gamma_1) = O(P_s)$. Therefore, $p_1 = O(P_s^{1/[1+\lambda]})$ and $z_2^\star = \min\{p_1, z_r\} = \min\{O(P_s^{1/[1+\lambda]}), z_r\} = z_r$. When $\lambda = 1$, $z_2^\star = \min\{\sqrt{(q_s\gamma_1+1)/(\gamma_2\gamma_3)}, z_r\} = \min\{O(P_s^{1/2}), z_r\} = z_r$. Thus, full power at both nodes, $q_1^\star = q_s$ and $z_2^\star = z_r$, is asymptotically optimal for $0 \leq \lambda \leq 1$ and the capacity $C_{\text{FD}} \to \log(1 + z_r\gamma_2)$ approaches a constant.

7.3.2 *Large Relay Power, Fixed Source Power*

Consider now the per-node constrained case where $P_r \to \infty$ while P_s remains fixed. When $\lambda = 0$, $z_2^\star = z_r$ and $C_{\text{FD}} \to \log(1 + [q_s\gamma_1]/[1+\gamma_3])$. When $0 < \lambda < 1$, applying the method of dominant balance to $P_1(z_2, \lambda)$, $O(z_2^{1+\lambda}\gamma_2\gamma_3) = O(z_2^\lambda\gamma_3) = O(q_s\gamma_1 + 1) = O(1)$. Therefore, $z_2^\star = p_1 = O(P_r^{-1})$ and the capacity also approaches a constant. When $\lambda = 1$, $z_2^\star = \sqrt{(q_s\gamma_1 + 1)/(\gamma_2\gamma_3)} = O(P_r^{-1})$ and the capacity approaches a constant as well. Note that when $\lambda > 0$, $z_2^\star P_r = O(1)$. Hence, as the power constraint at R approaches infinity, the power transmitted by the relay saturates to a certain value to control the self-interference.

7.3.3 *Large Source and Relay Power*

Consider the individual constraint case with $P_s = P_r \to \infty$. When $\lambda = 0$, $q_1^\star = q_s$, $z_2^\star = z_r$ and

$$C_{\text{FD}} = \log\left(\frac{q_1^\star z_2^\star\gamma_1\gamma_2 + O(P_s)}{q_1^\star\gamma_1 + z_2^\star\gamma_2 + z_2^\star\gamma_2\gamma_3 + O(1)}\right) = \log(P_s) + O(1).$$

The capacity can then be written as $C_{\text{FD}} = m \cdot \log(P_s) + O(1)$ where $m = 1$ is the multiplexing gain. When $0 < \lambda < 1$, applying the dominant balance method, $O(z_2^{1+\lambda}\gamma_2\gamma_3) = O(q_s\gamma_1) = O(P_s)$. The root $p_1 = z_2^\star = O(P_s^{-\lambda/[1+\lambda]})$ and

$$C_{\text{FD}} = \log\left(\frac{q_1^\star z_2^\star\gamma_1\gamma_2 + O(P_s)}{q_1^\star\gamma_1 + (z_2^\star)^{1+\lambda}\gamma_2\gamma_3 + O(P_s^{1/[1+\lambda]})}\right) = \frac{1}{1+\lambda}\log(P_s) + O(1),$$

with the multiplexing gain $m = 1/[1+\lambda]$. When $\lambda = 1$, $q_1^\star = q_s$, $z_2^\star = O(P_s^{-1/2})$ and

$$C_{\text{FD}} = \log\left(\frac{q_1^\star z_2^\star\gamma_1\gamma_2 + O(P_s)}{q_1^\star\gamma_1 + (z_2^\star)^2\gamma_2\gamma_3 + O(P_s^{1/2})}\right) = \frac{1}{2}\log(P_s) + O(1),$$

with $m = 1/2$. Note that the power transmitted by R, $z_2^\star P_s$, only grows as $O(P_s^{1/[1+\lambda]})$ when $\lambda > 0$, i.e., full relay power is not optimal.

7.3.4 *Large Global Power*

Finally, consider the joint constraint with $P_t \to \infty$. Following a similar approach as above, when $\lambda = 0$, $q_1^\star = O(1)$, $z_2^\star = O(1)$, and $C_{\text{FD}} = \log(P_t) + O(1)$ with a multiplexing gain $m = 1$. When $0 < \lambda < 1$, $q_1^\star = p_2 = q_t - O(P_t^{-\lambda/[1+\lambda]})$ and

Table 7.1 Capacity and power allocation behavior in high power regions

Scenario	$q_1^\star P_s$	$z_2^\star P_r$	m	Preferred protocol
High P_s, fixed P_r	$O(P_s)$	$O(1)$	–	FD
High P_r, fixed P_s	$O(1)$	$\begin{cases} O(P_r)\ (\lambda = 0) \\ O(1)\ \ (\lambda > 0) \end{cases}$	–	HD ($\lambda > 0$)
High $P_s = P_r$	$O(P_s)$	$O(P_r^{1/[1+\lambda]})$	$1/[1+\lambda]$	FD ($\lambda < 1$)
High P_t	$O(P_t)$	$O(P_t^{1/[1+\lambda]})$	$1/[1+\lambda]$	FD ($\lambda < 1$)

$z_2^\star = O(P_t^{-\lambda/[1+\lambda]})$. The capacity is then given by $C_{\text{FD}} = [1/(1+\lambda)]\log(P_t) + O(1)$ with $m = 1/[1+\lambda]$. When $\lambda = 1$, $A_1 > 0$, $q_1^\star = q_t - O(P_t^{-1/2})$ and $z_2^\star = O(P_t^{-1/2})$. Hence, $C_{\text{FD}} = [1/2]\log(P_t) + O(1)$ and $m = 1/2$.

The behavior of the capacity and optimal power allocation schemes for the above asymptotic cases is summarized in Table 7.1.

7.4 Illustrative Examples and Comparisons to HD Schemes

In this section, we provide simulation results to verify our theoretical analysis. Comparisons between the FD and HD schemes are also carried. For all simulations, unit noise power is considered $N_r = N_d = N_0 = 1$. To concentrate on the effect of λ, we set $\beta = 1$ in (7.1). In addition, to keep the same average power constraints between HD and FD, $q_s = z_r = q_t = 1$ for the FD system and $q_{s,\text{HD}} = z_{r,\text{HD}} = q_{t,\text{HD}} = 2$ for the HD one. Besides the optimal allocations derived in (7.4) and (7.5), the full ($q_1 = q_s$, $z_2 = z_r$) and uniform ($q_1 = z_2 = q_t/2$) schemes are also considered. Note that full power is optimal for HD with per-node constraints, whereas uniform allocation is optimal for HD under joint constraints with $\gamma_1 = \gamma_2$. Also note that the rate of the HD system can be obtained by setting $\gamma_3 = 0$ in (7.2) and pre-multiplying by a factor of $1/2$, i.e., the mutual information in (2.11).

Figure 7.1 shows the achievable rates of the DH system against P_s/N_0 for a fixed P_r/N_0 =5 dB and different values of λ. The performance of the ideal FD system without self-interference ($V = 0$) and using the optimal power scheme is also plotted as a benchmark. First, observe that full power is asymptotically optimal for the FD systems, as expected from Sect. 7.3. When compared to the HD system with full power, it can be seen that the FD systems perform better in high source power regions. This is because in this case, $C_{\text{FD}} \to \log(1 + z_r\gamma_2) > C_{\text{HD}} \to [1/2]\log(1+2z_r\gamma_2)$. Specifically, the FD systems provide an asymptotic 1.4× gain.

Figure 7.2 shows the achievable rates against P_r/N_0 for a fixed P_s/N_0 =5 dB. First, note that the proposed allocation schemes in (7.4) outperform full power for all values of λ. As discussed in Sect. 7.3, Fig. 7.2 indicates that using full power at the relay hurts the rate in high relay power regions when $\lambda > 0$. In addition, it can be seen that the HD system outperforms all FD schemes. This is because when $\lambda = 0$, it can be shown that $C_{\mathrm{HD}} \to [1/2]\log(1+2q_s\gamma_1) > C_{\mathrm{FD}} \to \log(1+[q_s\gamma_1]/[1+\gamma_3])$ when $2\gamma_3(1+\gamma_3) > q_s\gamma_1$, which is the case for the parameters in Fig. 7.2.

The rates of the DH system are plotted in Fig. 7.3 against $P_s/N_0 = P_r/N_0$. The rates in Fig. 7.3 are normalized by $\log(1+P_s)$. As before, note that the normalized rates of the FD systems using the allocations from Sect. 7.2 outperform those with full power. Moreover, the normalized rates with the optimal allocation approach the multiplexing gain values of $1/[1+\lambda]$, which agrees with Sect. 7.3. As $C_{\mathrm{HD}} = [1/2]\log(P_s) + O(1)$, the FD systems with the optimal allocation present a higher multiplexing gain than HD when $\lambda < 1$. In particular, the FD systems using the optimal strategy with $\lambda = [0, 0.3, 0.5, 0.7]$ provide gains of 1.6, 1.3, 1.2 and 1.0× at P_s/N_0 =20 dB. However, the capacity of the FD system with $\lambda = 1$ is

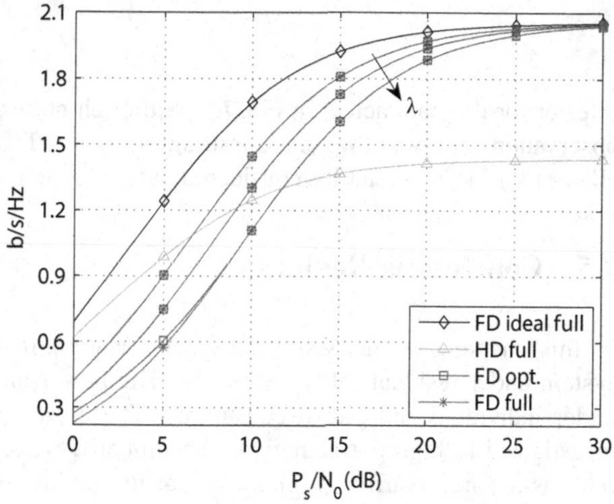

Fig. 7.1 Achievable rates against P_s/N_0 ($\alpha_1 = \alpha_2 = 1$, $\lambda = [0, 0.5, 1]$, P_r/N_0 =5 dB)

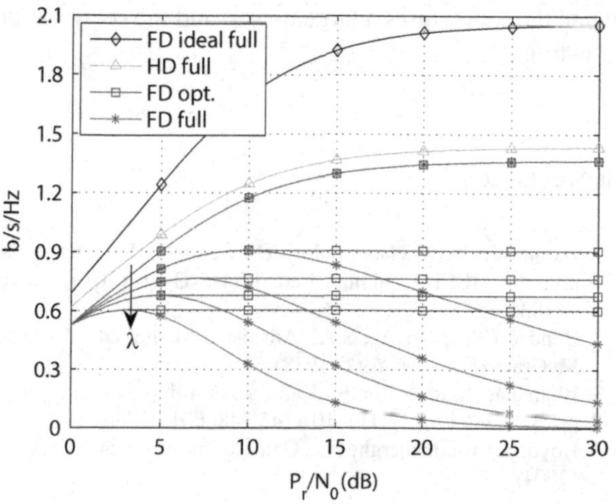

Fig. 7.2 Achievable rates against P_r/N_0 ($\alpha_1 = \alpha_2 = 1$, $\lambda = [0, 0.3, 0.5, 0.7, 1]$, P_s/N_0 =5 dB)

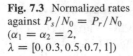

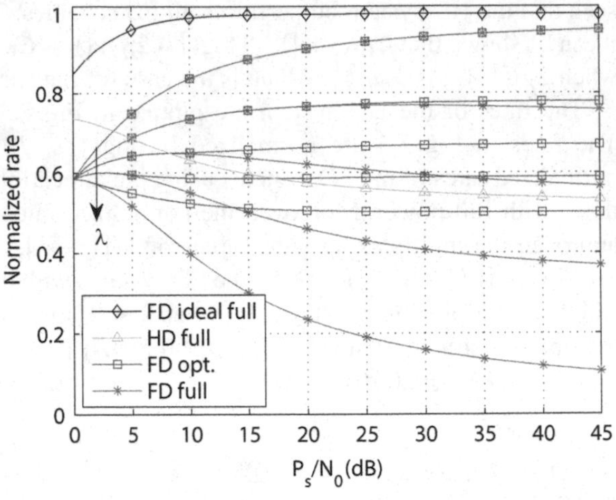

Fig. 7.3 Normalized rates against $P_s/N_0 = P_r/N_0$ ($\alpha_1 = \alpha_2 = 2$, $\lambda = [0, 0.3, 0.5, 0.7, 1]$)

inferior for the parameters in Fig. 7.3. Although not explicitly shown here, similar observations apply to the joint constraint scenario. The dominant protocols (either HD or FD) for the scenarios considered in this section are summarized in Table 7.1.

7.5 Concluding Remarks

In this chapter, we derived the capacity and optimal allocation for a FD DH system under residual self-interference. The power allocations were first obtained under individual and global constraints. The capacity and allocations were then investigated in high power regions. Specifically, we showed that full power at the relay is not necessarily optimal and that the multiplexing gain is $1/[1 + \lambda]$. Our analysis also revealed that the FD system is able to outperform its HD counterpart when the source uses a large power, and the relay has either a fixed or a large power constraint.

References

1. Baranwal, T., Michalopoulos, D., Schober, R.: Outage analysis of multihop full-duplex relaying. IEEE Commun. Lett. **17**(1), 63–66 (2013). Doi:10.1109/LCOMM.2012.112812.121826
2. Bender, C.M., Orszag, S.A.: Advanced Mathematical Methods for Scientists and Engineers. McGraw-Hill, New York (1978)
3. Bharadia, D., McMilin, E., Katti, S.: Full-duplex radios. In: Proceeding of ACM SIGCOMM, pp. 375–386 (2013). Doi:10.1145/2486001.2486033
4. Boyd, S., Vandenberghe, L.: Convex Optimization. Cambridge University Press, Cambridge (2004)

5. Day, B., Margetts, A., Bliss, D., Schniter, P.: Full-duplex MIMO relaying: Achievable rates under limited dynamic range. IEEE J. Sel. Areas Commun. **30**(8), 1541–1553 (2012). Doi:10.1109/JSAC.2012.120921
6. Duarte, M., Dick, C., Sabharwal, A.: Experiment-driven characterization of full-duplex wireless systems. IEEE Trans. Wirel. Commun. **11**(12), 4296–4307 (2012). Doi:10.1109/TWC.2012.102612.111278
7. Jain, M., Choi, J.I., Kim, T., Bharadia, D., Seth, S., Srinivasan, K., Levis, P., Katti, S., Sinha, P.: Practical, real-time, full-duplex wireless. In: Proceeding of ACM Annual International Conference Mobile Computing Network, pp. 301–312 (2011). Doi:10.1145/2030613.2030647
8. Jiménez-Rodríguez, L., Tran, N.H., Le-Ngoc, T.: Optimal power allocation and capacity of full-duplex AF relaying under residual self-interference. IEEE Wirel. Commun. Lett. **3**(2), 233–236 (2014). Doi:10.1109/WCL.2014.020614.130831
9. Jingmei, Z., Qi, Z., Chunju, S., Ying, W., Ping, Z., Zhang, Z.: Adaptive optimal transmit power allocation for two-hop non-regenerative wireless relaying system. In: Proceeding of IEEE Vehicular Technology, vol. 2, pp. 1213–1217 (2004). Doi:10.1109/VETECS.2004.1389025
10. Kim, T.M., Paulraj, A.: Outage probability of amplify-and-forward cooperation with full-duplex relay. In: Proceedings of IEEE Wireless Communications and Networking Conference, pp. 75–79 (2012). Doi:10.1109/WCNC.2012.6214473
11. Krikidis, I., Suraweera, H., Smith, P., Yuen, C.: Full-duplex relay selection for amplify-and-forward cooperative networks. IEEE Trans. Wirel. Commun. **11**(12), 4381–4393 (2012). Doi:10.1109/TWC.2012.101912.111944
12. Krikidis, I., Suraweera, H., Yang, S., Berberidis, K.: Full-duplex relaying over block fading channel: A diversity perspective. IEEE Trans. Wirel. Commun. **11**(12), 4524–4535 (2012). Doi:10.1109/TWC.2012.102612.120254
13. Michalopoulos, D., Schlenker, J., Cheng, J., Schober, R.: Error rate analysis of full-duplex relaying. In: Proceedings of Internayional Waveform Diversity Design Conference, pp. 165–168 (2010). Doi:10.1109/WDD.2010.5592409
14. Ng, D., Lo, E., Schober, R.: Dynamic resource allocation in MIMO-OFDMA systems with full-duplex and hybrid relaying. IEEE Trans. Commun. **60**(5), 1291–1304 (2012). Doi:10.1109/TCOMM.2012.031712.110233
15. Riihonen, T., Werner, S., Wichman, R.: Optimized gain control for single-frequency relaying with loop interference. IEEE Trans. Wirel. Commun. **8**(6), 2801–2806 (2009). Doi:10.1109/TWC.2009.080542
16. Riihonen, T., Werner, S., Wichman, R.: Hybrid full-duplex/half-duplex relaying with transmit power adaptation. IEEE Trans. Wirel. Commun. **10**(9), 3074–3085 (2011). Doi:10.1109/TWC.2011.071411.102266
17. Riihonen, T., Werner, S., Wichman, R.: Mitigation of loopback self-interference in full-duplex MIMO relays. IEEE Trans. Signal Process. **59**(12), 5983–5993 (2011). Doi:10.1109/TSP.2011.2164910
18. Riihonen, T., Werner, S., Wichman, R., Eduardo, Z.: On the feasibility of full-duplex relaying in the presence of loop interference. In: Proceeding of IEEE Signal Processing Advances in Wireless Communications, pp. 275–279 (2009). Doi:10.1109/SPAWC.2009.5161790
19. Riihonen, T., Werner, S., Wichman, R., Hamalainen, J.: Outage probabilities in infrastructure-based single-frequency relay links. In: Wireless Communications and Networking, IEEE Conference, pp. 1–6 (2009). Doi:10.1109/WCNC.2009.4917875
20. Shomorony, I., Avestimehr, A.: Is Gaussian noise the worst-case additive noise in wireless networks? In: Proceeding of IEEE International Symposium on Information Theory, pp. 214–218 (2012). Doi:10.1109/ISIT.2012.6283743
21. Suraweera, H., Krikidis, I., Zheng, G., Yuen, C., Smith, P.: Low-complexity end-to-end performance optimization in MIMO full-duplex relay systems. IEEE Trans. Wirel. Commun. **13**(2), 913–927 (2014). Doi:10.1109/TWC.2013.122313.130608

Chapter 8
Full-Duplex AF Relaying: Error Performance Under Residual Self-Interference

In Chap. 7, we studied the capacity of FD relaying assuming Gaussian codebooks at the source. We now consider the error performance of FD systems using finite constellations. As explained in Chap. 7, most works on FD relaying with residual self-interference assume that the variance of the interference is proportional to the average transmitted power. Moreover, studies in the literature have only considered the non-cooperative dual-hop approach, in which the direct link from source to destination is either assumed to be a source of interference [1, 4, 14, 16, 18], or completely ignored [12, 13, 15–17, 19]. While these non-cooperative FD schemes allow the source to transmit continuously, their error performance has been shown to be significantly affected in the above residual self-interference model. For example, it has been shown via BER and outage analysis that the FD systems present a *zero* diversity order [1, 12, 14, 19], i.e., the outage or BER curves present an error floor. Such floor persists even with relay [12] or antenna [19] selection.

As noted in Chap. 7, recent experimental results in [2, 7] suggest that the variance of the residual self-interference is proportional to the λ-th power of the average transmitted power, where λ is less than one. This means that previous works have only considered the worst-case scenario of $\lambda = 1$. In addition, we know from the studies on cooperative HD protocols that the direct link is important in terms of both rate and diversity advantages. Therefore, it is expected that similar benefits can be exploited in FD systems. However, to our knowledge, the idea of cooperative relaying has not been explored in FD systems under self-interference.

Inspired by the above observations, this chapter investigates the error and diversity performance of FD AF relay systems under the residual self-interference model in [7]. We consider the cooperative LR system that makes use of the direct link [5, 8], and the DH system without direct link (both introduced in Sect. 2.3). We shall examine the error and diversity performances of the uncoded system, and the coded system under the framework of BICM similar to Chap. 6. As a first step, we derive closed-form expressions of the uncoded PEP for the LR and DH protocols.

L. Jiménez Rodríguez et al., *Amplify-and-Forward Relaying in Wireless Communications*, SpringerBriefs in Computer Science,
DOI 10.1007/978-3-319-17981-0_8

The derived expressions can be used not only to analyze the uncoded systems, but also to obtain tight bounds to the BER of the coded ones. Asymptotic PEP and BER expressions in high transmission power regions are then obtained. Based on these expressions, it is shown that FD LR systems can achieve the same diversity function as their HD counterparts [6, 9, 10] for any $0 \leq \lambda \leq 1$ as long as a suitable precoder at the source is applied. Different from previous works where the direct link is treated as interference, a non-zero diversity order is attained and no error floor is observed despite the self-interference in FD. By further analyzing the coding gain of the LR system when $\lambda > 0$, it is demonstrated that the source only needs to transmit a superposition of L signals in the first time slot to maximize this gain. Equivalently, it turns out that using HD orthogonal relaying together with a superposition constellation is asymptotically optimal. For DH systems, the asymptotic analysis indicates that their diversity functions are dependent on λ. In particular, the diversity order is a decreasing function of λ for $0 < \lambda \leq 1$ and an error floor is observed only when $\lambda = 1$. The DH system must then sacrifice its diversity when $\lambda > 0$ to transmit in FD mode. Although HD relaying is asymptotically optimal for both relaying protocols, illustrative results show that FD relaying is advantageous at practical BER levels in good self-interference cancellation scenarios.

8.1 System Model

The general block diagram of the uncoded relay systems is shown in Fig. 8.1. First, for the uncoded FD LR protocol, the information sequence $\underline{\boldsymbol{u}}$ is divided into groups of Lm_c bits. Each group is mapped to a signal $\boldsymbol{s} = [s_1, \ldots, s_L]^\top$ in the L-D constellation Ψ. Each component s_i is assumed to be in a 1-D unit-energy constellation Ω of size 2^{m_c}. We assume that the mapping ξ from binary bits to symbols is implemented *independently* in Ω for each component s_i. The symbol $\boldsymbol{s}$ is then rotated by a $L \times L$ single-frame precoder as $\boldsymbol{x} = \boldsymbol{G}\boldsymbol{s}$. As we demonstrate shortly and similar to Chap. 6, the use of such precoder provides diversity benefits for the LR system. Each precoded symbol $\boldsymbol{x}$ is then sent over L periods according to the LR protocol in (2.18) with $\alpha_l > 0$. At the destination, a maximum likelihood (ML) detector is applied on the received vector $\boldsymbol{y}$ to obtain the estimated information sequence $\underline{\breve{\boldsymbol{u}}}$.

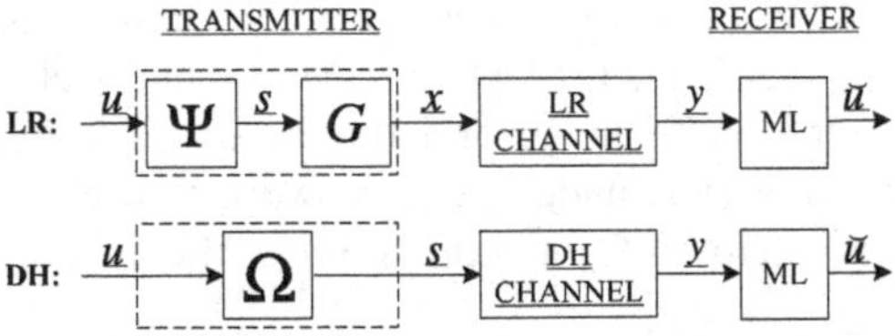

Fig. 8.1 Uncoded LR and DH systems

For the uncoded FD DH protocol, $\underline{u}$ is divided into groups of m_c bits which are mapped directly to $x_i = s_i \in \Omega$ using the mapping rule ξ. Given that space diversity is not exploited in DH relaying, no precoder is considered in this case. The signals x_i are then transmitted using the DH protocol in (2.24), where the direct link is assumed to be heavily shadowed with $\alpha_0 = 0$ similar to Chap. 7. At the destination, ML decoding is applied on the received signal y_i as in the LR case.

In the coded framework, the general block diagram of the LR and DH BICM-ID systems is similar to the one described in Fig. 6.1. Specifically, the information sequence $\underline{u}$ in Fig. 8.1 is first encoded and then interleaved prior to modulation. At the destination, as illustrated in Fig. 6.1, the received sequence is decoded in an iterative manner using a MAP demodulator and a SISO channel decoder.

In this chapter, we consider the fading environment in which the transmitted codeword spans numerous channel realizations of $\boldsymbol{h} = [h_0, h_1, h_2]$. Moreover, the destination possesses full CSI, whereas the relay has only statistical knowledge of the incoming S-R link. Assuming that the relay only amplifies the signal received in the previous period, the FG amplification coefficients for the LR and DH protocols are given by (2.22) and (2.25) with $\Xi = \phi$, respectively. Recall that the covariance matrices of the signals transmitted by S and R in the LR protocol are given by $\boldsymbol{Q} = \mathbb{E}[\boldsymbol{x}\boldsymbol{x}^{\dagger}] = \boldsymbol{G}\boldsymbol{G}^{\dagger}$ and $\boldsymbol{Z} = \mathbb{E}[\boldsymbol{t}\boldsymbol{t}^{\dagger}]$. We assume that $\mathrm{tr}(\boldsymbol{Q}) = L$ and $\mathrm{tr}(\boldsymbol{Z}) = L - 1$ so that the source and relay have respectively an average power constraint of P_s and P_r per symbol period when active. The entries of $\boldsymbol{G}$, $\{g_{i,k}\}$ $(1 \leq i, k \leq L)$, must then satisfy the power constraint $\mathrm{tr}(\boldsymbol{Q}) = \mathrm{tr}(\boldsymbol{G}\boldsymbol{G}^{\dagger}) = \sum_{i=1}^{L}\sum_{k=1}^{L} |g_{i,k}|^2 = L$. In a similar manner, we assume that $\mathbb{E}[|s_i|^2] = 1$ and $z_2 = 1$ for the DH protocol.

Finally, similar to Chap. 7, we adopt the residual self-interference model in [7]. Specifically, the residual self-interference is modeled as a complex Gaussian random variable with variance given by

$$V = \beta P_r^{\lambda}, \tag{8.1}$$

where P_r is the average power transmitted by the relay when active, and β and λ $(0 \leq \lambda \leq 1)$ are constants that can be found in [2, 7].

8.2 PEP Analysis and Tight BER Bounds

In this section, we first derive closed-form PEP expressions for the uncoded systems. Such derivation is also useful in developing tight BER bounds for the coded ones.

8.2.1 PEP for the LR System

The PEP is defined as the probability of deciding in favor of $\breve{s} = [\breve{s}_1, \ldots, \breve{s}_L]^\top$ given that s was transmitted, $\breve{s}, s \in \Psi$ and $s \neq \breve{s}$. These two signal points correspond to the rotated symbols $x = Gs$ and $\breve{x} = [\breve{x}_1, \ldots, \breve{x}_L]^\top = G\breve{s}$. Assuming perfect CSI at D, the PEP conditioned on h can be written from (2.18) as

$$\mathbb{P}(s \to \breve{s}|h) = Q\left(\sqrt{\frac{d^2(s, \breve{s}|h)}{2}}\right), \tag{8.2}$$

where $Q(\cdot)$ is the Gaussian Q-function and $d^2(s, \breve{s}|h)$ is the squared Euclidean distance conditioned on h:

$$d^2(s, \breve{s}|h) = P_s\|K^{-1/2}H_{\mathrm{LR}}(x - \breve{x})\|^2 = P_s\|K^{-1/2}H_{\mathrm{LR}}G(s - \breve{s})\|^2. \tag{8.3}$$

Following a similar approach as in Chap. 6 and assuming that only the previous symbol is forwarded by the relay, the signal component in (2.18) can be alternatively written as

$$H_{\mathrm{LR}}Gs = H_{\mathrm{LR}}x = XTh_{01}, \tag{8.4}$$

where the 2×1 $h_{01} = [h_0, h_1]^\top$, the $L \times 2$

$$T = \begin{pmatrix} 1 & 0 \\ 0 & \sqrt{P_r}h_2b_{2,1} \\ \vdots & \vdots \\ 0 & \sqrt{P_r}h_2b_{L,L-1} \end{pmatrix}, \tag{8.5}$$

and the $L \times L$

$$X = \begin{pmatrix} x_1 & 0 & \cdots & \cdots & \cdots & 0 \\ x_2 & x_1 & 0 & \cdots & \cdots & 0 \\ x_3 & 0 & x_2 & 0 & \cdots & 0 \\ \vdots & \vdots & \ddots & \ddots & \ddots & \vdots \\ \vdots & \vdots & \ddots & \ddots & \ddots & 0 \\ x_L & 0 & \cdots & \cdots & 0 & x_{L-1} \end{pmatrix}. \tag{8.6}$$

The distance in (8.3) can then be expressed from (8.4) as

$$d^2(s, \breve{s}|h) = P_s h_{01}^\dagger T^\dagger U^\dagger K^{-1} U T h_{01}, \tag{8.7}$$

where $\boldsymbol{U} = \boldsymbol{X} - \breve{\boldsymbol{X}}$ and $\breve{\boldsymbol{X}}$ is given as in (8.6) but replacing x_i by $\breve{x}_i$. Applying the alternate representation of the Gaussian integral $Q(x) = \frac{1}{\pi}\int_0^{\pi/2} \exp\left(-x^2/[2\sin^2\theta]\right) d\theta$ and from (8.7), the conditional PEP becomes

$$\mathbb{P}(\boldsymbol{s} \to \breve{\boldsymbol{s}}|\boldsymbol{h}) = \frac{1}{\pi}\int_0^{\pi/2} \exp\left(-c_\theta \boldsymbol{h}_{01}^\dagger \boldsymbol{T}^\dagger \boldsymbol{U}^\dagger \boldsymbol{K}^{-1} \boldsymbol{U}\boldsymbol{T}\boldsymbol{h}_{01}\right) d\theta,$$

where $c_\theta = P_s/[4\sin^2\theta]$. Averaging over $\boldsymbol{h}$, the unconditional PEP can be written as

$$\mathbb{P}(\boldsymbol{s} \to \breve{\boldsymbol{s}}) = \frac{1}{\pi}\int_0^{\pi/2} \Delta_\theta(\boldsymbol{s}, \breve{\boldsymbol{s}})\, d\theta, \tag{8.8}$$

where

$$\Delta_\theta(\boldsymbol{s}, \breve{\boldsymbol{s}}) = \mathbb{E}_{\boldsymbol{h}}\left[\exp\left(-c_\theta \boldsymbol{h}_{01}^\dagger \boldsymbol{T}^\dagger \boldsymbol{U}^\dagger \boldsymbol{K}^{-1} \boldsymbol{U}\boldsymbol{T}\boldsymbol{h}_{01}\right)\right]. \tag{8.9}$$

To solve the expectation in (8.9), we can first use the fact that given $\boldsymbol{a} \sim \mathscr{CN}(\boldsymbol{0}, \boldsymbol{\Phi})$ and a Hermitian matrix $\boldsymbol{A}$, $\mathbb{E}[\exp(-\boldsymbol{a}^H \boldsymbol{A}\boldsymbol{a})] = 1/\det(\boldsymbol{I} + \boldsymbol{\Phi}\boldsymbol{A})$ [6]. Thus, we have

$$\Delta_\theta(\boldsymbol{s}, \breve{\boldsymbol{s}}) = \mathbb{E}_{h_2}\left[\frac{1}{\det(\boldsymbol{I}_2 + c_\theta \boldsymbol{\Phi}_{01} \boldsymbol{T}^\dagger \boldsymbol{U}^\dagger \boldsymbol{K}^{-1} \boldsymbol{U}\boldsymbol{T})}\right], \tag{8.10}$$

where $\boldsymbol{\Phi}_{01} = \begin{pmatrix} \phi_0 & 0 \\ 0 & \phi_1 \end{pmatrix}$. Given that the relay only forwards the previous symbol, the covariance in (2.20) is written as $\boldsymbol{K} = \mathrm{diag}(\boldsymbol{k})$ with the $L \times 1$ vector $\boldsymbol{k}$:

$$\begin{aligned}\boldsymbol{k} = [k_1, \ldots, k_L]^\top =& [N_d, N_d + N_r P_r \alpha_2 |b_{2,1}|^2, N_d + P_r \alpha_2 |b_{3,2}|^2 (V + N_r), \\ & \ldots, N_d + P_r \alpha_2 |b_{L,L-1}|^2 (V + N_r)]^\top. \end{aligned} \tag{8.11}$$

Let the second column of $\boldsymbol{U}\boldsymbol{T}$ be given by

$$\boldsymbol{\tau} = [\tau_1, \ldots, \tau_L]^\top = [0, \sqrt{P_r} h_2 b_{2,1} u_1, \ldots, \sqrt{P_r} h_2 b_{L,L-1} u_{L-1}]^\top,$$

with $\boldsymbol{u} = [u_1, \ldots, u_L] = \boldsymbol{x} - \breve{\boldsymbol{x}}$. The determinant in (8.10) can then be written as

$$\begin{aligned}\det(\boldsymbol{I}_2 + c_\theta \boldsymbol{\Phi}_{01} \boldsymbol{T}^\dagger \boldsymbol{U}^\dagger \boldsymbol{K}^{-1} \boldsymbol{U}\boldsymbol{T}) &= \det\begin{pmatrix} 1 + c_\theta \phi_0 \sum_{l=1}^{L} \frac{|u_l|^2}{k_l} & c_\theta \phi_0 \sum_{l=2}^{L} \frac{\tau_l u_l^\dagger}{k_l} \\ c_\theta \phi_1 \sum_{l=2}^{L} \frac{\tau_l^\dagger u_l}{k_l} & 1 + c_\theta \phi_1 \sum_{l=2}^{L} \frac{|\tau_l|^2}{k_l} \end{pmatrix} \\ &= P_1(\alpha_2)/P_2(\alpha_2),\end{aligned}$$

where

$$
\begin{aligned}
P_1(\alpha_2) =& c_\theta\phi_0 \left(\sum_{l=1}^{L} |u_l|^2 \prod_{\substack{m=1\\ m\neq l}}^{L} k_m \right) + c_\theta\phi_1[k_1 + c_\theta\phi_0|u_1|^2] \left(\sum_{l=2}^{L} |\tau_l|^2 \prod_{\substack{m=2\\ m\neq l}}^{L} k_m \right) \\
&+ \left(\prod_{l=1}^{L} k_l \right) + c_\theta^2\phi_0\phi_1 k_1 \left(\sum_{l=2}^{L} \sum_{\substack{m=2\\ m\neq l}}^{L} [|u_l|^2|\tau_m|^2 - \tau_l^\dagger u_l \tau_m u_m^\dagger] \prod_{\substack{n=2\\ n\neq l\\ n\neq m}}^{L} k_n \right),
\end{aligned}
$$

$$
P_2(\alpha_2) = \prod_{l=1}^{L} k_l .
$$

It can be shown from the definition of $\boldsymbol{\tau}$ and $\boldsymbol{k}$ that the functions $P_1(\alpha_2)$ and $P_2(\alpha_2)$ are polynomials in α_2 of order $L-1$. Let p_l be the roots of $P_1(\alpha_2)$ $(1 \leq l \leq L-1)$. Using partial fraction expansion,

$$
\frac{P_2(\alpha_2)}{P_1(\alpha_2)} = \omega_0 + \sum_{l=1}^{L-1} \frac{\omega_l}{\alpha_2 - p_l},
$$

where ω_l are constants that depend on $\boldsymbol{\tau}$ and $\boldsymbol{k}$, and can be obtained from the partial fraction procedure for a given L. The expectation in (8.9) can finally be solved as

$$
\Delta_\theta(\boldsymbol{s}, \check{\boldsymbol{s}}) = \mathbb{E}_{\alpha_2}\left[\frac{P_2(\alpha_2)}{P_1(\alpha_2)}\right] = \omega_0 + \sum_{l=1}^{L-1} \mathbb{E}_{\alpha_2}\left[\frac{\omega_l}{\alpha_2 - p_l}\right] = \omega_0 + \frac{1}{\phi_2} \sum_{l=1}^{L-1} \omega_l \mathscr{J}\left(\frac{-p_l}{\phi_2}\right), \tag{8.12}
$$

with $\mathscr{J}(x) = \exp(x) E_1(x)$ and the exponential integral $E_1(x)$ in (4.4). The PEP for the LR system can then be obtained from (8.8) by substituting $\Delta_\theta(\cdot, \cdot)$ as in (8.12).

8.2.2 *PEP for the DH System*

We now consider the PEP for the DH protocol with $h_0 = 0$. In the DH case, the conditional PEP $\mathbb{P}(s_i \to \check{s}_i | h_1, h_2)$ can be written as in (8.2) with

$$
d^2(s_i, \check{s}_i | h_1, h_2) = \frac{P_s P_r \alpha_1 \alpha_2 b^2 |\epsilon_i|^2}{P_r \alpha_2 b^2 (N_r + V) + N_d},
$$

and $\epsilon_i = s_i - \breve{s}_i \neq 0$. Following a similar approach as in (8.8), the unconditional PEP is given by

$$\mathbb{P}(s_i \to \breve{s}_i) = \frac{1}{\pi}\int_0^{\pi/2} \Delta_\theta(s_i, \breve{s}_i)\, d\theta, \tag{8.13}$$

where

$$\Delta_\theta(s_i, \breve{s}_i) = \mathbb{E}_{h_1,h_2}\left[\exp\left(\frac{-P_s P_r \alpha_1 \alpha_2 b^2 |\epsilon_i|^2}{4\sin^2\theta \cdot [P_r \alpha_2 b^2 (N_r + V) + N_d]}\right)\right]. \tag{8.14}$$

The expectation in (8.14) can then be solved similar to (8.10) as:

$$\Delta_\theta(s_i, \breve{s}_i) = \mathbb{E}_{h_2}\left[\left(1 + \frac{P_s P_r \phi_1 \alpha_2 b^2 |\epsilon_i|^2}{4\sin^2\theta \cdot [P_r \alpha_2 b^2 (N_r + V) + N_d]}\right)^{-1}\right] = \mathbb{E}_{h_2}\left[\frac{A_2\alpha_2 + B_2}{C_2\alpha_2 + B_2}\right],$$

where $A_2 = 4\sin^2\theta \cdot P_r b^2 (N_r + V)$, $B_2 = 4\sin^2\theta \cdot N_d$ and $C_2 = P_r b^2 [4\sin^2\theta \cdot (N_r + V) + P_s |\epsilon_i|^2 \phi_1]$. Using partial fraction expansion,

$$\begin{aligned} \Delta_\theta(s_i, \breve{s}_i) &= \frac{A_2}{C_2} + \frac{B_2 C_2 - B_2 A_2}{C_2^2} \mathbb{E}_{h_2}\left[\frac{1}{\alpha_2 + (B_2/C_2)}\right] \\ &= \frac{A_2}{C_2} + \frac{B_2 C_2 - B_2 A_2}{\phi_2 C_2^2} \mathscr{I}\left(\frac{B_2}{\phi_2 C_2}\right). \end{aligned} \tag{8.15}$$

The PEP for the DH system is then obtained by substituting (8.15) in (8.13).

8.2.3 Tight BER Bounds of BICM-ID Systems

The PEP expressions in (8.12) and (8.15) can be used not only to analyze uncoded systems, but also to provide tight bounds to the BER of the coded systems as shown in Chap. 6. In particular, by applying the error-free feedback approach, the union bound on the BER of a LR BICM-ID system that employs a rate-k_c/n_c code can be written as in (6.4) with $f(d, \Psi, \xi, \boldsymbol{G})$ upper bounded as in (6.11) but now with

$$\gamma_\theta(\Psi, \xi, \boldsymbol{G}) = \frac{1}{Lm_c 2^{Lm_c}} \sum_{\boldsymbol{s}\in\Psi} \sum_{k=1}^{Lm_c} \Delta_\theta(\boldsymbol{s}, \boldsymbol{p}). \tag{8.16}$$

The bound to the BER of the coded LR system is then obtained by substituting (8.12), (6.11) and (8.16) into (6.4). A similar approach as in (6.4) can be followed for the coded DH BICM-ID system using (8.15). As will be shown in Sect. 8.4, the bound in (6.4) is tight at sufficiently high transmit powers for both LR and DH protocols. Hence, this bound is useful to predict the BER performance of the coded relay systems without resorting to lengthy Monte Carlo simulations.

8.3 Diversity and Coding Gain Analysis

In the previous section, closed-form expressions of the uncoded PEP were derived for the LR and DH protocols, which were also useful for the development of tight bounds to the error performance of coded systems. However, the derived expressions do not provide insights about the diversity performance of these systems. In this section, focusing on the high power region, we develop simplified PEP expressions that are useful to carry out the diversity and coding gain analysis under uncoded and coded frameworks. For simplicity, hereafter, it is assumed that $P_s = P_r$.

8.3.1 LR Analysis

For the LR protocol, we first consider the diversity analysis for the uncoded system before extending the results to the coded one. Insights into the precoder design shall also be discussed.

8.3.1.1 Uncoded System

For the uncoded LR system, the PEP in (8.8) can first be upper bounded using the Chernoff bound $Q(\sqrt{2x}) < [1/2]\exp(-x)$ as

$$\mathbb{P}(\boldsymbol{s} \to \breve{\boldsymbol{s}}) = \frac{1}{\pi}\int_0^{\pi/2} \Delta_\theta(\boldsymbol{s}, \breve{\boldsymbol{s}})\, d\theta < \frac{1}{2}\Delta_\theta(\boldsymbol{s}, \breve{\boldsymbol{s}})\Big|_{\theta=\pi/2}, \tag{8.17}$$

where $\Delta_\theta(\cdot,\cdot)$ is given in (8.12). Let $\boldsymbol{\epsilon} = \boldsymbol{s} - \breve{\boldsymbol{s}} = [\epsilon_1, \ldots, \epsilon_L] \neq \boldsymbol{0}$ and $\boldsymbol{g}_i^{\mathsf{T}}$ be the i-th row of $\boldsymbol{G}$. As $\Delta_\theta(\cdot,\cdot)$ is still very involved even when $\theta = \pi/2$, we have the following proposition with regards to this function in high transmit power regions.

Proposition 8.1. *When P_s is sufficiently large, $\Delta_{\pi/2}(\cdot,\cdot)$ in (8.12) simplifies to*

$$\Delta_{\pi/2}(\boldsymbol{s}, \breve{\boldsymbol{s}}) = \begin{cases} \frac{16N_d^2\eta}{\phi_0\phi_2(L-1)\mathcal{U}} \cdot P_s^{-2}\ln(P_s) + O(P_s^{-2}), & \lambda = 0 \\ \frac{16N_d^2\lambda\eta}{\phi_0\phi_2(L-1)^2|u_1|^4} \cdot P_s^{-2}\ln(P_s) + O(P_s^{-2}), & 0 < \lambda < 1 \\ \frac{16N_d^2[\eta\phi_1+(L-2)\beta]}{\phi_0\phi_1\phi_2(L-1)^2|u_1|^4} \cdot P_s^{-2}\ln(P_s) + O(P_s^{-2}), & \lambda = 1. \end{cases} \tag{8.18}$$

In (8.18), $u_i = \boldsymbol{g}_i^{\top}\boldsymbol{\epsilon}$, $\mathscr{U} = (\sum_{i=1}^{L}|u_i|^2)(\sum_{i=1}^{L-1}|u_i|^2) - |\sum_{i=1}^{L-1}u_i u_{i+1}^{\dagger}|^2$ *and* $\eta = \sum_{i=1}^{L-1} q_i = \sum_{i=1}^{L-1}\sum_{k=1}^{L}|g_{i,k}|^2$.

Proof. The proof follows by first ignoring the lower order terms of $P_1(\cdot)$ and $P_2(\cdot)$, and then carrying the expectation of the remaining terms [11]. □

Substituting (8.18) into (8.17) and ignoring the lower order terms, the PEP for the uncoded LR system can be approximated for high transmitted powers as

$$\mathbb{P}(\boldsymbol{s} \to \breve{\boldsymbol{s}}) \approx \frac{8N_d^2 \cdot \mathscr{F}_{\text{unc}}^{\text{LR}}(\boldsymbol{G}, \boldsymbol{\epsilon})}{\phi_0\phi_2(L-1)} \cdot P_s^{-2}\ln(P_s), \tag{8.19}$$

where

$$\mathscr{F}_{\text{unc}}^{\text{LR}}(\boldsymbol{G}, \boldsymbol{\epsilon}) = \begin{cases} \eta/\mathscr{U}, & \lambda = 0 \\ \frac{\lambda\eta}{(L-1)|u_1|^4}, & 0 < \lambda < 1 \\ \frac{\eta\phi_1 + (L-2)\beta}{\phi_1(L-1)|u_1|^4}, & \lambda = 1. \end{cases} \tag{8.20}$$

First, observe from (8.19) that a full-diversity gain function of $P_s^{-2}\ln(P_s)$ can be achieved for any $0 \le \lambda \le 1$ as long as the coding gain function $\mathscr{F}_{\text{unc}}^{\text{LR}}(\boldsymbol{G}, \boldsymbol{\epsilon})^{-1} \neq 0$. This diversity function is the same as the one obtained for the uncoded HD NAF system studied in [6, 9]. Thus, any precoder with $|u_1| = |\boldsymbol{g}_1^{\top}\boldsymbol{\epsilon}| \neq 0$ for $0 < \lambda \le 1$ or $\mathscr{U} \neq 0$ for $\lambda = 0$ offers a full-diversity performance for the uncoded LR system. More importantly, thanks to the use of the direct S-D link, the FD LR system does not suffer from an error floor as long as a full-diversity precoder is used.

Among the class of full-diversity $\boldsymbol{G}$, it is then of interest to find the optimal precoder that maximizes the coding gain using the worst-case PEP analysis. Let $\eta_i = \|\boldsymbol{g}_i\|^2$. From (8.19), this is equivalent to solving the following:

$$\min_{\boldsymbol{G}} \max_{\boldsymbol{\epsilon} \neq \boldsymbol{0}} \; \mathscr{F}_{\text{unc}}^{\text{LR}}(\boldsymbol{G}, \boldsymbol{\epsilon}) \quad \text{s.t.} \quad \sum_{i=1}^{L}\eta_i = L. \tag{8.21}$$

Consider first the case of $0 < \lambda \le 1$. As an initial step, we find the optimal value for the magnitude of each row of $\boldsymbol{G}$, $\eta_i^{\star}$. Let $\boldsymbol{g}_i = \sqrt{\eta_i}\tilde{\boldsymbol{g}}_i = \sqrt{\eta_i}\cdot[\tilde{g}_{i,1}, \ldots, \tilde{g}_{i,L}]^{\top}$, where $\|\tilde{\boldsymbol{g}}_i\|^2 = 1$. Then, $\eta/|u_1|^4$ in (8.20) for $0 < \lambda < 1$ can be written as

$$\frac{\eta}{|u_1|^4} = \frac{\sum_{l=1}^{L-1}\eta_l}{\eta_1^2|\tilde{\boldsymbol{g}}_1^{\top}\boldsymbol{\epsilon}|^4} = \frac{1}{|\tilde{\boldsymbol{g}}_1^{\top}\boldsymbol{\epsilon}|^4}\left(\frac{1}{\eta_1} + \frac{\sum_{l=2}^{L-1}\eta_l}{\eta_1^2}\right).$$

The optimal values of η_i for $0 < \lambda < 1$ can then be found by minimizing $[1/\eta_1] + \sum_{l=2}^{L-1}[\eta_l/\eta_1^2]$. It is easy to see that this term is decreasing with η_1 and increasing with η_l for $2 \le l \le L-1$. Since $\sum_{l=1}^{L}\eta_l = L$, at optimality, $\eta_1^{\star} = L$ and $\eta_l^{\star} = 0$

for $2 \leq l \leq L$. For $\lambda = 1$, $1/[\eta_1^2|\tilde{\boldsymbol{g}}_1^{\mathsf{T}}\boldsymbol{\epsilon}|^4]$ is also decreasing with η_1. Therefore, $\eta_1^\star = L$ and $\eta_l^\star = 0$ for $2 < l \leq L$ are optimal. The problem in (8.21) simplifies to

$$\max_{\boldsymbol{g}_1} \min_{\boldsymbol{\epsilon} \neq \boldsymbol{0}} |u_1|^2 \quad \text{s.t.} \quad \|\boldsymbol{g}_1\|^2 = L. \tag{8.22}$$

From the above derivation, the optimal strategy for $0 < \lambda \leq 1$ is to transmit from the source a superposition of all symbols in the first time slot ($i = 1$) using all the power and to remain silent in the following $L - 1$ time slots. The relay in turn must amplify and forward in the second slot $i = 2$ the information it received in the previous time. Hence, at sufficiently large P_s, the relay is in fact trying to transmit in HD mode to maximize the coding gain. Consequently, given that the source and relay transmit in an orthogonal manner, HD orthogonal relaying using a superposition constellation can be considered asymptotically optimal. A similar behavior was observed in [9] for the uncoded HD NAF system.

Note that the above observations hold as long as $V = O(P_s^\lambda)$ with $\lambda > 0$. When $\lambda = 0$, $V = O(1)$ and the asymptotic performance depends not only on u_1 and η, but also on u_i for $2 \leq i \leq L$ through the parameter $\mathscr{U}$. Therefore, an explicit construction of the optimal precoder appears to be more challenging and is left for future work.

8.3.1.2 Coded System

Similar to the uncoded system, it is straightforward to show that the BER of the coded LR scheme can be approximated by considering the most significant term in (6.4) and applying the Chernoff bound to (6.11) as

$$P_{\mathrm{b}} \approx \frac{c_{d_{\mathrm{H}}}}{2k_{\mathrm{c}}} \left[\gamma_\theta(\Psi, \xi, \boldsymbol{G})\right]^{d_{\mathrm{H}}} \bigg|_{\theta=\pi/2}. \tag{8.23}$$

Then, substituting the simplified expression of (8.18) in (8.16) and ignoring the lower order terms, $\gamma_{\pi/2}(\Psi, \xi, \boldsymbol{G})$ can be approximated as

$$\gamma_{\pi/2}(\Psi, \xi, \boldsymbol{G}) \approx \frac{16 N_d^2 P_s^{-2} \ln(P_s)}{L m_{\mathrm{c}} 2^{L m_{\mathrm{c}}} \phi_0 \phi_2 (L-1)} \sum_{\boldsymbol{s} \in \Psi} \sum_{k=1}^{L m_{\mathrm{c}}} \mathscr{F}_{\mathrm{unc}}^{\mathrm{LR}}(\boldsymbol{G}, \boldsymbol{\epsilon}).$$

When the labeling ξ is implemented independently and identically for each component $s_i \in \Omega$, $\boldsymbol{\epsilon} = \boldsymbol{s} - \boldsymbol{p}$ has only one non-zero component. The above function can then be simplified as in Chap. 6 by averaging over the 1-bit neighbors $s, p \in \Omega$ with $\epsilon = s - p \neq 0$ (rather than $\boldsymbol{s}, \boldsymbol{p} \in \Psi$) as

$$\gamma_{\pi/2}(\Psi, \xi, \boldsymbol{G}) = \frac{16 N_d^2 P_s^{-2} \ln(P_s)}{L m_{\mathrm{c}} 2^{m_{\mathrm{c}}} \phi_0 \phi_2 (L-1)} \cdot \mathscr{F}_{\mathrm{code}}^{\mathrm{LR}}(\boldsymbol{G}) \cdot \sum_{s \in \Omega} \sum_{k=1}^{m_{\mathrm{c}}} \frac{1}{|\epsilon|^4},$$

with

$$\mathscr{F}_{\mathrm{code}}^{\mathrm{LR}}(\boldsymbol{G}) \tag{8.24}$$
$$= \begin{cases} \eta \sum_{l=1}^{L} \left[\left(\sum_{i=1}^{L} |g_{i,l}|^2 \right) \left(\sum_{i=1}^{L-1} |g_{i,l}|^2 \right) - \left| \sum_{i=1}^{L-1} g_{i,l} g_{i+1,l}^{\dagger} \right|^2 \right]^{-1}, & \lambda = 0 \\ \frac{\lambda}{L-1} \left(\sum_{l=1}^{L-1} \frac{\eta_l}{\eta_1^2} \right) \left(\sum_{l=1}^{L} \frac{1}{|\tilde{g}_{1,l}|^4} \right), & 0 < \lambda < 1 \\ \frac{1}{L-1} \left(\frac{(L-2)\beta}{\phi_1 \eta_1^2} + \sum_{l=1}^{L-1} \frac{\eta_l}{\eta_1^2} \right) \left(\sum_{l=1}^{L} \frac{1}{|\tilde{g}_{1,l}|^4} \right), & \lambda = 1. \end{cases}$$

Then, the BER expression in (8.23) simplifies to

$$P_{\mathrm{b}} \approx \frac{c_{d_{\mathrm{H}}}}{2k_{\mathrm{c}}} \left[\frac{16 N_d^2 \cdot \mathscr{F}_{\mathrm{code}}^{\mathrm{LR}}(\boldsymbol{G})}{L m_{\mathrm{c}} 2^{m_{\mathrm{c}}} \phi_0 \phi_2 (L-1)} \cdot \sum_{s \in \Omega} \sum_{k=1}^{m_{\mathrm{c}}} \frac{1}{|\epsilon|^4} \right]^{d_{\mathrm{H}}} \times [P_s^{-2} \ln(P_s)]^{d_{\mathrm{H}}}. \tag{8.25}$$

Observe from (8.25) that a full-diversity gain function of $[P_s^{-2}\ln(P_s)]^{d_{\mathrm{H}}}$ can be achieved for any $0 \le \lambda \le 1$ as long as the coding gain function $\mathscr{F}_{\mathrm{code}}^{\mathrm{LR}}(\boldsymbol{G})^{-1} \neq 0$. As in the uncoded case, this is the same diversity gain function as the one observed for the coded HD NAF system in [10], and any precoder with $\mathscr{F}_{\mathrm{code}}^{\mathrm{LR}}(\boldsymbol{G})^{-1} \neq 0$ belongs to the class of full-diversity precoders for the coded LR system. Similarly, the coded system does not present an error floor as long as a full-diversity precoder is used.

The optimal precoder $\boldsymbol{G}^{\star}$ for the coded system can then be obtained by minimizing (8.25), or equivalently (8.24), given $\sum_{i=1}^{L} \eta_i = L$. As before, by focusing on $0 < \lambda \le 1$, it can be easily shown that $\mathscr{F}_{\mathrm{code}}^{\mathrm{LR}}(\boldsymbol{G})$ is a decreasing with η_1 and an increasing with η_l for $2 \le l \le L-1$. Thus, $\eta_1^{\star} = L$ and $\eta_l^{\star} = 0$ for $2 \le l \le L$. Furthermore, by setting η_i to the above values, it can be shown from the arithmetic/geometric mean inequality that $|g_{1,k}^{\star}|^2 = 1$ for $1 \le k \le L$ minimizes $\mathscr{F}_{\mathrm{code}}^{\mathrm{LR}}(\boldsymbol{G})$ when $\lambda > 0$. Therefore, as in the uncoded scenario, transmitting a superposition constellation using HD orthogonal relaying is asymptotically optimal for $0 < \lambda \le 1$. This behavior has also been observed for the coded HD NAF system in [10].

8.3.2 DH Analysis

We now turn our attention to the analysis of the uncoded and coded DH systems.

8.3.2.1 Uncoded System

First, the PEP of the uncoded DH system can be upper bounded similar to (8.17) by replacing $\Delta_\theta(\boldsymbol{s}, \check{\boldsymbol{s}})$ with $\Delta_\theta(s_i, \check{s}_i)$ in (8.15). As in the LR analysis, we then have the following proposition with regards to the asymptotic behavior of $\Delta_\theta(\cdot, \cdot)$ in (8.15).

Proposition 8.2. *When P_s is sufficiently large, $\Delta_{\pi/2}(\cdot, \cdot)$ in (8.15) simplifies to*

$$\Delta_{\pi/2}(s_i, \check{s}_i) = \begin{cases} \frac{4N_d}{\phi_2|\epsilon_i|^2} \cdot P_s^{-1}\ln(P_s) + O(P_s^{-1}), & \lambda = 0 \\ \frac{4\beta}{\phi_1|\epsilon_i|^2} \cdot P_s^{-(1-\lambda)} + O(P_s^{-1}\ln(P_s)), & 0 < \lambda < 1 \\ \frac{4\beta}{4\beta + \phi_1|\epsilon_i|^2} + O(P_s^{-1}\ln(P_s)), & \lambda = 1. \end{cases} \tag{8.26}$$

Proof. The proof follows by ignoring the lower order terms of $\Delta_{\pi/2}(\cdot, \cdot)$ in (8.15) (please see [11] for details). □

Based on the above proposition and (8.17), the PEP for the uncoded DH system can be approximated as

$$\mathbb{P}(s_i \to \check{s}_i) \approx 2 \cdot \mathscr{F}_{\text{unc}}^{\text{DH}}(\epsilon_i) \cdot \mathscr{D}_{\text{unc}}^{\text{DH}}(P_s), \tag{8.27}$$

where

$$\mathscr{F}_{\text{unc}}^{\text{DH}}(\epsilon_i) = \begin{cases} N_d/[\phi_2|\epsilon_i|^2], & \lambda = 0 \\ \beta/[\phi_1|\epsilon_i|^2], & 0 < \lambda < 1 \\ \beta/[4\beta + \phi_1|\epsilon_i|^2], & \lambda = 1, \end{cases} \tag{8.28}$$

and the diversity gain function

$$\mathscr{D}_{\text{unc}}^{\text{DH}}(P_s) = \begin{cases} P_s^{-1}\ln(P_s), & \lambda = 0 \\ P_s^{-(1-\lambda)}, & 0 < \lambda < 1 \\ 1, & \lambda = 1. \end{cases} \tag{8.29}$$

First, observe from (8.27) and (8.29) that different from the LR system, the diversity gain function of the DH scheme depends on λ. In particular, a diversity gain function of $P_s^{-(1-\lambda)}$ is attained when $0 < \lambda < 1$, whereas the gain function is $P_s^{-1}\ln(P_s)$ when $\lambda = 0$. Similar to previous works, an error floor is observed when $\lambda = 1$. Following the same approach as in Proposition 8.2 for $\lambda = 0$, it can be easily shown that the diversity function of a HD DH system is $P_s^{-1}\ln(P_s)$. Thus, in contrast to the HD scheme, the FD DH system can achieve a full rate of one transmitted symbol per period but must sacrifice its diversity to $P_s^{-(1-\lambda)}$ when $\lambda > 0$.

8.3.2.2 Coded System

For the coded DH system, the BER can be approximated following the approach in (8.23) as

$$P_{\mathrm{b}} \approx \frac{c_{d_{\mathrm{H}}}}{2k_{\mathrm{c}}} \left[\frac{4}{m_c 2^{m_c}} \sum_{s \in \Omega} \sum_{k=1}^{m_{\mathrm{c}}} \mathscr{F}_{\mathrm{unc}}^{\mathrm{DH}}(\epsilon) \right]^{d_{\mathrm{H}}} \times [\mathscr{D}_{\mathrm{unc}}^{\mathrm{DH}}(P_s)]^{d_{\mathrm{H}}}. \tag{8.30}$$

Similar to the uncoded case, observe from (8.30) that the diversity function depends on the value of λ. The diversity function is $P_s^{-(1-\lambda)d_{\mathrm{H}}}$ for $0 < \lambda < 1$ and $[P_s^{-1} \ln(P_s)]^{d_{\mathrm{H}}}$ for $\lambda = 0$. A zero-diversity order is still attained when $\lambda = 1$.

8.4 Illustrative Examples

In this section, we provide simulation results to verify the theoretical analysis carried out in previous sections. In all simulations, the Gray-labeled QPSK constellation is employed unless otherwise stated. The channel gains are assumed to be Rayleigh-distributed with unit variance. The powers at S and R are also assumed to be equal $P_s = P_r$ and unit noise power at both nodes is considered $N_r = N_d = N_0 = 1$. Furthermore, to concentrate on the effect on λ, we set $\beta = 1$ in (8.1). For the coded systems, the rate-1/2 convolutional code with generator matrix $\{5; 7\}$ is considered along with an 80,000-length bit interleaver.

8.4.1 LR Examples

Figure 8.2 shows the BER performance of the uncoded LR system for $L = 3$ against bit SNR E_b/N_0 (in dB). The 3×3 precoders considered in Fig. 8.2 are the identity $\boldsymbol{I}_3$ precoder (i.e., no precoder) and the $\boldsymbol{G}_{\mathrm{cyclo3}}$ precoder from [20]. Furthermore, three values of λ in (8.1) are considered: $\lambda = 0, 0.5$, and 1. First, it can be checked from (8.20) that $\boldsymbol{G}_{\mathrm{cyclo3}}$ is a full-diversity precoder for the uncoded system. As a consequence, the slopes of the BER curves with $\boldsymbol{G}_{\mathrm{cyclo3}}$ remain unchanged for any λ and match with the diversity order $P_s^{-2} \log(P_s)$, which is also plotted in Fig. 8.2 as a reference. On the other hand, $\boldsymbol{I}_3$ is not a full-diversity precoder and thus it can be seen from Fig. 8.2 that its diversity performance depends on the value of λ. Specifically, an error floor is observed when $\lambda = 1$.

To demonstrate the asymptotic optimality of HD relaying, Fig. 8.2 also depicts the BER performance of the uncoded LR system using the optimal 3×3 precoder $\boldsymbol{G}_{\mathrm{ex3}}$, where $\boldsymbol{g}_{1,\mathrm{ex3}}^{\top} = [0.755, 0.387e^{j0.236}, 1.510]$ and $\boldsymbol{g}_{i,\mathrm{ex3}}^{\top} = \boldsymbol{0}$ for $i = 2, 3$. The vector $\boldsymbol{g}_{1,\mathrm{ex3}}$ is the solution to (8.22) for the QPSK constellation and was obtained using exhaustive search. Note that since the optimal precoder transforms the FD

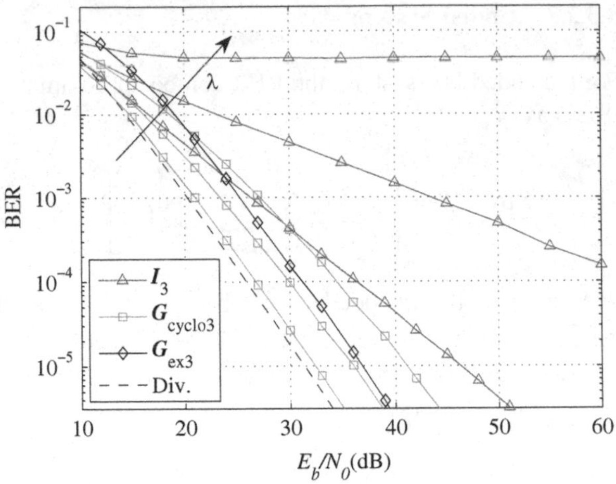

Fig. 8.2 BER performances of uncoded LR using different 3×3 precoders ($\lambda = 0, 0.5, 1$)

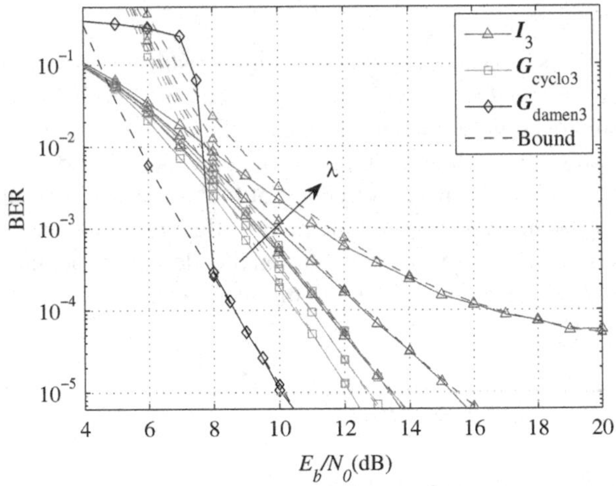

Fig. 8.3 BER performances of LR BICM-ID using different 3×3 precoders ($\lambda = 0, 0.5, 1$)

system into a HD one, its performance is independent of λ. Observe from Fig. 8.2 that at the BER level of 10^{-5}, the $\boldsymbol{G}_{\text{ex3}}$ precoder outperforms $\boldsymbol{G}_{\text{cyclo3}}$ when $\lambda > 0.5$. However, at this BER level, it is yet not able to outperform $\boldsymbol{G}_{\text{cyclo3}}$ with $\lambda = 0.5$ and the crossover happens only at the BER of 10^{-6}. Therefore, although HD transmission is asymptotically optimal, FD might be better at practical BER levels when $\lambda \leq 0.5$.

The diversity behavior in the coded scenario is illustrated in Fig. 8.3 for $L = 3$ and $\lambda = \{0, 0.5, 1\}$. In particular, Fig. 8.3 shows the BER performances of the LR BICM-ID system after ten decoding iterations along with the BER bound in (6.4). First, observe from Fig. 8.3 that the BER performance of all systems converges to the bound in (6.4) at a sufficiently high transmitted power. Specifically, the analytical and simulation results converge around the practical BER levels of $10^{-3} - 10^{-4}$. This confirms the accuracy of the PEP expressions derived in (8.8) and (8.12). Moreover, it can be checked from (8.24) that $\boldsymbol{G}_{\text{cyclo3}}$ is a full-diversity precoder for the coded LR system, whereas $\boldsymbol{I}_3$ is not. Hence, the BER curves with $\boldsymbol{G}_{\text{cyclo3}}$ present the same slope, while the slopes of the identity systems change with λ. An error floor can still be observed in Fig. 8.3 when $\lambda = 1$. This is in agreement with Sect. 8.3.1.

Figure 8.3 also shows the BER performance of the coded LR system using the $\boldsymbol{G}_{\text{damen3}}$ precoder with $\boldsymbol{g}_{1,\text{damen3}}^{\top} = [1, e^{j\varphi}, e^{j2\varphi}]$ and $\boldsymbol{g}_{i,\text{damen3}}^{\top} = \boldsymbol{0}$ for $i = 2, 3$. This precoder is asymptotically optimal for the coded system as it minimizes (8.24). Furthermore, the angle is set to $\varphi = 0.312$, which is the solution to (8.22) for the structure $\boldsymbol{g}_1^{\top} = [1, e^{j\varphi}, e^{j2\varphi}]$ [3]. Such angle guarantees a faster convergence to the bound [10]. It can be seen from Fig. 8.3 that the asymptotic performance of $\boldsymbol{G}_{\text{damen3}}$ outperforms those of $\boldsymbol{G}_{\text{cyclo3}}$ for any value of λ. In particular, the LR system with $\boldsymbol{G}_{\text{damen3}}$ presents a gain of more than 2 dB over $\boldsymbol{G}_{\text{cyclo3}}$ at the BER level of 10^{-5}. Thus, different from the uncoded case, the HD system takes advantage of the iterative gain in the BICM-ID structure and it is preferred over FD at this practical BER level.

8.4.2 DH Examples

The BER performances of the uncoded DH system are shown in Fig. 8.4 for different values of λ. The diversity orders in (8.29) are also illustrated in Fig. 8.4. First, note that the slopes of the BER curves are in agreement with the diversity analysis in Sect. 8.3.2. As expected, it can then be seen from Fig. 8.4 that the diversity behavior is highly dependent on the value of λ. In particular, the diversity order is a decreasing function of $\lambda > 0$ and an error floor is observed in Fig. 8.4 when $\lambda = 1$.

Similar trends for the coded scenario can be seen in Fig. 8.5. For the DH BICM system, it was observed that iterations do not significantly improve the performance.

Fig. 8.4 BER performances of the uncoded DH system for different values of λ

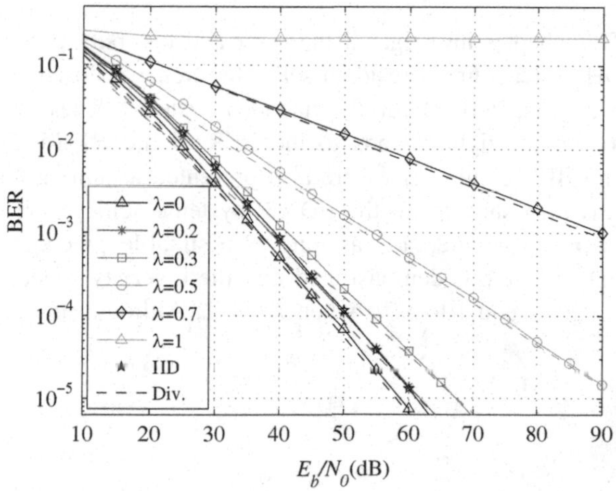

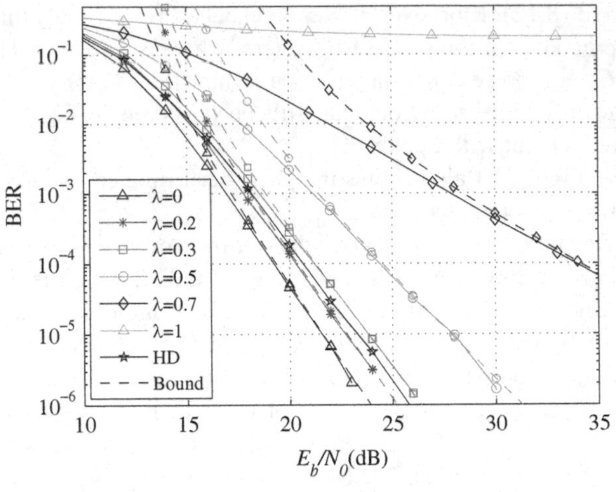

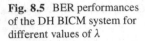

Fig. 8.5 BER performances of the DH BICM system for different values of λ

As such, all curves in Fig. 8.5 are plotted after 1 iteration. First, note that the BER bounds are tight for all considered systems. This confirms the correctness of the expressions derived in (8.13) and (8.15). As in the uncoded scenario, it can be seen from Fig. 8.5 that λ has a great impact on the diversity order of the coded system.

Finally, to compare the FD and HD schemes, the BER of the uncoded/coded HD DH systems is also plotted in Figs. 8.4 and 8.5. To maintain the same throughput, the HD systems employ the Gray-labeled 16-QAM modulation scheme. Note from these figures that at the BER level of 10^{-5}, the HD systems outperform all FD schemes with $\lambda \geq 0.3$. However, the FD systems are advantageous when $\lambda \leq 0.2$. As such, for practical purposes, $\lambda < 0.3$ for FD to be worth using in this scenario.

8.5 Concluding Remarks

This chapter investigated the error and diversity performance of the FD LR and DH systems under residual self-interference. Closed-form expressions of the PEP were first derived for the uncoded systems. These expressions were then used to provide tight bounds to the BER of the BICM-ID systems. Simplified PEP and BER expressions were also presented assuming high transmission powers. It was then shown that the FD LR systems achieve the same diversity function as their HD counterparts as long as a suitable precoder is applied. In the case of FD DH, it was demonstrated that the diversity order is a decreasing function of λ and is equal to zero when $\lambda = 1$. Although the HD systems were shown to be asymptotically optimal, simulations results revealed that their FD counterparts might be advantageous at practical BER levels when a high cancellation quality is used, i.e., when λ is small.

References

1. Baranwal, T., Michalopoulos, D., Schober, R.: Outage analysis of multihop full-duplex relaying. IEEE Commun. Lett. **17**(1), 63–66 (2013). Doi:10.1109/LCOMM.2012.112812.121826
2. Bharadia, D., McMilin, E., Katti, S.: Full-duplex radios. In: Proceedings of ACM SIGCOMM, pp. 375–386 (2013). Doi:10.1145/2486001.2486033
3. Damen, M.: Joint coding/decoding in a multiple access system, application to mobile communications. Ph.D. thesis, ENST de Paris, France (1999)
4. Day, B., Margetts, A., Bliss, D., Schniter, P.: Full-duplex MIMO relaying: Achievable rates under limited dynamic range. IEEE J. Sel. Areas Commun. **30**(8), 1541–1553 (2012). Doi:10.1109/JSAC.2012.120921
5. Del Coso, A., Ibars, C.: Achievable rates for the AWGN channel with multiple parallel relays. IEEE Trans. Wirel. Commun. **8**(5), 2524–2534 (2009). Doi:10.1109/TWC.2009.080288
6. Ding, Y., Zhang, J.K., Wong, K.M.: The amplify-and-forward half-duplex cooperative system: Pairwise error probability and precoder design. IEEE Trans. Signal Process. **55**(2), 605–617 (2007). Doi:10.1109/TSP.2006.885761
7. Duarte, M., Dick, C., Sabharwal, A.: Experiment-driven characterization of full-duplex wireless systems. IEEE Trans. Wirel. Commun. **11**(12), 4296–4307 (2012). Doi:10.1109/TWC.2012.102612.111278
8. El Gamal, A., Mohseni, M., Zahedi, S.: Bounds on capacity and minimum energy-per-bit for AWGN relay channels. IEEE Trans. Inf. Theory **52**(4), 1545–1561 (2006). Doi:10.1109/TIT.2006.871579
9. Jiménez-Rodríguez, L., Tran, N.H., Le-Ngoc, T.: Jointly optimal precoder and power allocation for an amplify-and-forward half-duplex relay system. In: Proceedings of IEEE International Conference on Advanced Technologies for Communications, pp. 305–310. HCM City, Vietnam (2010). Doi:10.1109/ATC.2010.5672703
10. Jiménez-Rodríguez, L., Tran, N.H., Le-Ngoc, T.: Bandwidth-efficient bit-interleaved coded modulation over NAF relay channels: Error performance and precoder design. IEEE Trans. Veh. Technol. **60**(5), 2086–2101 (2011). Doi:10.1109/TVT.2011.2138176
11. Jiménez-Rodríguez, L., Tran, N.H., Le-Ngoc, T.: Performance of full-duplex AF relaying in the presence of residual self-interference. IEEE J. Sel. Areas Commun. **32**(9), 1752–1764 (2014). Doi:10.1109/JSAC.2014.2330151
12. Krikidis, I., Suraweera, H., Smith, P., Yuen, C.: Full-duplex relay selection for amplify-and-forward cooperative networks. IEEE Trans. Wirel. Commun. **11**(12), 4381–4393 (2012). Doi:10.1109/TWC.2012.101912.111944
13. Krikidis, I., Suraweera, H., Yang, S., Berberidis, K.: Full-duplex relaying over block fading channel: A diversity perspective. IEEE Trans. Wirel. Commun. **11**(12), 4524–4535 (2012). Doi:10.1109/TWC.2012.102612.120254
14. Michalopoulos, D., Schlenker, J., Cheng, J., Schober, R.: Error rate analysis of full-duplex relaying. In: Proceeding of International Waveform Diversity Design Conference, pp. 165–168 (2010). Doi:10.1109/WDD.2010.5592409
15. Ng, D., Lo, E., Schober, R.: Dynamic resource allocation in MIMO-OFDMA systems with full-duplex and hybrid relaying. IEEE Trans. Commun. **60**(5), 1291–1304 (2012). Doi:10.1109/TCOMM.2012.031712.110233

16. Riihonen, T., Werner, S., Wichman, R.: Hybrid full-duplex/half-duplex relaying with transmit power adaptation. IEEE Trans. Wirel. Commun. **10**(9), 3074–3085 (2011). Doi:10.1109/TWC.2011.071411.102266
17. Riihonen, T., Werner, S., Wichman, R., Eduardo, Z.: On the feasibility of full-duplex relaying in the presence of loop interference. In: Proceeding of IEEE Signal Processing Advances in Wireless Communications, pp. 275–279 (2009). Doi:10.1109/SPAWC.2009.5161790
18. Riihonen, T., Werner, S., Wichman, R., Hamalainen, J.: Outage probabilities in infrastructure-based single-frequency relay links. In: Proceeding of IEEE Wireless Communications and Networking Conference, pp. 1–6 (2009). Doi:10.1109/WCNC.2009.4917875
19. Suraweera, H., Krikidis, I., Zheng, G., Yuen, C., Smith, P.: Low-complexity end-to-end performance optimization in MIMO full-duplex relay systems. IEEE Trans. Wirel. Commun. **13**(2), 913–927 (2014). Doi:10.1109/TWC.2013.122313.130608
20. Viterbo, E.: Full diversity rotations. URL http://www.ecse.monash.edu.au/staff/eviterbo/rotations/rotations.html

Printed in the United States
By Bookmasters